APPLIED PRINCIPLES OF HYDROLOGY

JOHN C. MANNING
Professor Emeritus
California State College–Bakersfield

Illustrated by Natalie J. Weiskal

CBS

CBS PUBLISHERS & DISTRIBUTORS PVT. LTD.

New Delhi • Bengaluru • Chennai • Kochi • Mumbai • Pune

ISBN: 978-81-239-1588-3

First Indian Edition: 1989

Sales Area: India only

Published by:
Satish Kumar Jain for CBS Publishers & Distributors Pvt. Ltd.,
4819/XI Prahlad Street, 24 Ansari Road, Daryaganj, New Delhi - 110002
delhi@cbspd.com, cbspubs@airtelmail.in • www.cbspd.com
Ph.: 23289259, 23266861, 23266867 • Fax: 011-23243014

Corporate Office: 204 FIE, Industrial Area, Patparganj, Delhi - 110 092
Ph: 49344934 • Fax: 011-49344935
E-mail: publishing@cbspd.com • publicity@cbspd.com

Branches:
- ***Bengaluru:*** 2975, 17th Cross, K.R. Road, Bansankari 2nd Stage, Bengaluru - 70 • Ph: +91-80-26771678/79 • Fax: +91-80-26771680 E-mail: cbsbng@gmail.com, bangalore@cbspd.com
- ***Chennai:*** No. 7, Subbaraya Street, Shenoy Nagar, Chennai - 600030 Ph: +91-44-26681266, 26680620 • Fax: +91-44-42032115 E-mail: chennai@cbspd.com
- ***Kochi:*** 36/14, Kalluvilakam, Lissie Hospital Road, Kochi - 682018 Ph: +91-484-4059061-65 • Fax: +91-484-4059065 E-mail: cochin@cbspd.com
- ***Mumbai:*** 83-C, Dr. E. Moses Road, Worli, Mumbai - 400018 Ph: +91-9833017933, 022-24902340/41 • E-mail: mumbai@cbspd.com
- ***Pune:*** Bhuruk Prestige, Sr. No. 52/12/2+1+3/2, Narhe, Haveli (Near Katraj-Dehu Road Bypass), Pune - 411041 Ph: +91-20-64704058/59, 32342277 • E-mail: pune@cbspd.com

Representatives:
- Hyderabad: 0-9885175004
- Kolkata: 0-9831437309, 0-9051152362
- Nagpur: 0-9021734563
- Patna: 0-9334159340
- Vijayawada: 0-9000660880

Printed at: Nazia Printers, Delhi

Preface

Along with the air we breathe, water is probably the most used, the most abused, and the most taken-for-granted natural resource. Nearly everyone accepts as a matter of faith that water will always pour from a faucet when it is turned on. Few people realize how much effort has gone into planning and building the great urban water-distribution systems. Do you know, for example, where your city gets its water?

Years ago New York City built aqueducts to bring water from the Catskill Mountains; Los Angeles has one aqueduct that runs far up the eastern side of the Sierra Nevada Range and another that crosses the Mojave Desert to tap the Colorado River; Denver even gets some of its water from a tunnel that runs under the continental divide to the west slope of the Rocky Mountains; Chicago is lucky—it has Lake Michigan in its front yard.

Whatever the source, abundant supplies of good, clean water are essential for any community. In the United States a growing population and a rising standard of living have caused total water use to grow steadily during recent years. From 1950 to 1980 the population increased 1½ times from about 150 million to 230 million, whereas total withdrawals of water from all sources grew about 2½ times from 180 billion to 450 billion gallons per day (681 to 1703 million m^3/day).

Although supplies continue to be adequate in most parts of the United States, it has become clear that there are environmental costs connected with an ever-increasing draft on the nation's water resources. Efforts toward conservation and environmental protection have increased the need to know how water behaves as it moves through the water cycle. And the behavior of water in nature is what hydrology is all about.

This book presents a brief outline of hydrology, beginning with a discussion of the physical and chemical attributes that make water such a unique substance, and going on to a step-by-step discussion of the water cycle. Wherever possible the explanations of scientific principles are illustrated by examples of actual occurrences from the real world. Then at the end of each chapter is a section entitled "Applications." These are more complete accounts of real occurrences and are intended to show how to apply the principles of hydrology to practical problems in everyday life.

Material for this book came from class notes and from my own experience as a professional hydrologist and engineering geologist. From 1970 until 1980, I taught a course in elementary hydrology at the California State College in Bakersfield. Cal. State–Bakersfield opened in 1970 as the 19th campus of the State University and College System and was given the mission of specializing in environmental studies. Hydrology was included in the science curriculum, and I was asked to design a course that would be acceptable to undergraduates who were not majoring in science and who might not have an extensive background in mathematics.

Selecting a textbook for the course was difficult. Most books on water were either too technical or too generalized, and I experimented for several years with a variety of books and reading assignments from the library. In the end I gradually developed a set of lecture materials and class notes that seemed to do the job pretty well. After retiring from active teaching in 1980, I decided to work these up into a more coherent form, and the present text is the result.

As with any book on a technical subject, this one draws heavily on previously published works. However, because the book is not intended as a scientific treatise, no effort has been made to document text material with a list of cited references. Instead, a list of annotated references is included in Appendix IV for anyone who wishes to further pursue a topic mentioned in the text. Also because so many examples in the text are from California, a map with localities noted is included in an appendix.

Acknowledgements

I would like to express my thanks to friends and colleagues who gave help and encouragement both during the course on which this book is based and during preparation of the manuscript. I owe a special word of thanks to the following colleagues for material aid during preparation of the manuscript: J. P. Bluemle, North Dakota Geological Survey; G. Bogart, Water Dept., City of Bakersfield, California; P. Burdick, Valley Waste Disposal Co. (California); J. Goodridge, Consultant (California); D. R. Hetzel, Colorado River Commission of Nevada; R. L. Morrison, U.S. Bureau of Reclamation; N. A. MacGillivray, California Department of Water Resources; Professor L. J. Paulson, University of Nevada at Las Vegas; and C. E. Trotter, Arvin-Edison Water Storage District, (California).

Finally, I am especially grateful to Dr. John R. Coash, former Dean of the School of Arts and Science at Cal. State–Bakersfield, for his support and encouragement.

John C. Manning

Concerning Units

Although the English-speaking countries have been moving toward full adoption of the metric system, and no doubt will continue to do so, the United States has been generally indifferent to metric measure. U.S. consumers still buy milk in quart or half-gallon containers, although most containers also note the capacity in liters. Food is sold by the avoirdupois ounce or pound, gold and silver are traded in troy ounces and pounds, petroleum is measured in 42-gallon barrels, corn and soybeans are traded in bushels, and so on.

U.S. scientists have long used metric units, but engineers have only recently begun to convert their measurements; the transition is still far from complete. Most hydrologic measurements in the United States are still recorded and published in "American" units (they were known as English units before Great Britain officially adopted the metric system). For example, the U.S. Geological Survey Circular 1001, "Estimated Use of Water in the United States in 1980," published in 1983, reports water use in gallons and acre-feet. The National Weather Service reports rainfall and evaporation in inches; however, it regularly reports temperatures in both degrees Fahrenheit and Celsius.

In this text, most quantities have been converted to one or the other of the two systems. The quantity being discussed is stated first, as it was measured, and the conversion follows in parentheses. Most illustrations, however, show units in only one system.

It will probably be many years before all hydrologic data is published in metric units. Even then, hydrologists will still need to be conversant in both systems because the science requires interpreting current events, which nearly always depends on a study of past records. It is assumed that the reader has access to a table of metric conversions.

Contents

1

The Water Cycle

Everywhere water is a thing of beauty, gleaming in the dewdrop; singing in the summer rain; shining in the ice-gems till the leaves all seem to turn to living jewels; spreading a golden veil over the setting sun; or a white gauze around the midnight moon. **(J. B. Gough, "A Glass of Water")**

EARTH—THE BLUE PLANET

The Apollo voyages to the moon gave us the first view of Earth from deep space. The astronauts recorded some of these views in a series of strikingly beautiful photographs that show Earth as a small, blue planet floating in the darkness of space. The presence of water vapor in the sky and liquid water on the surface give the planet its bright blue color. After the monochromatic deadness of the moon, the returning astronauts undoubtedly were elated as they approached Earth and saw the white clouds and blue waters of their home planet.

Earth is the only planet in the solar system with liquid water on its surface, and therefore the only planet where life can originate and thrive. If there is water on Mars, it is in the form of ice; on Venus, because of high surface temperatures, water will occur only in the atmosphere as vapor. The large outer planets (Jupiter, Saturn, Uranus, and Neptune) are far too cold to contain liquid water. Earth's ability to support liquid water may result from its position in the solar system. Scientists have speculated that if Earth were only 6 million miles (9.6 million km) closer to the sun, it could have evolved in the same way as Venus and in so doing become uninhabitable.

Earth's orbital position yielded moderate temperatures and liquid water on its surface. Early in its history the atmosphere lost most of its carbon dioxide, which was absorbed by ocean water and subsequently locked up in carbonate sediments and sedimentary rocks (limestones). Minus the excess carbon dioxide, the atmosphere never developed the runaway greenhouse effect that developed on Venus, where the gas heated the surface, vaporizing all of the liquid water. Earth's more moderate surface temperatures support water in its three states: ice, liquid water, and vapor.

Unlike other planets, Earth also has the unique ability to supply water vapor to the atmosphere and fresh water to the land continuously. This process, called the water cycle, is as important to humankind as the very existence of liquid water itself. As long as the ocean exists as liquid water, life probably could exist there indefinitely. Without the water cycle, however, the land would be barren and lifeless.

THE WATER CYCLE

The water cycle, also called the hydrologic cycle, is the continuous circulation of water from the sea to the atmosphere to the land and back again to the sea. Its major components are shown in Figure 1.1. Table 1.1 indicates the approximate volumes of water passing annually through the cycle. A summary account of Earth's total water supply appears in Table 1.2.

Comparing figures in Table 1.1 and Table 1.2 shows what a small volume of the total water supply (about 1%) actually passes through the hydrologic

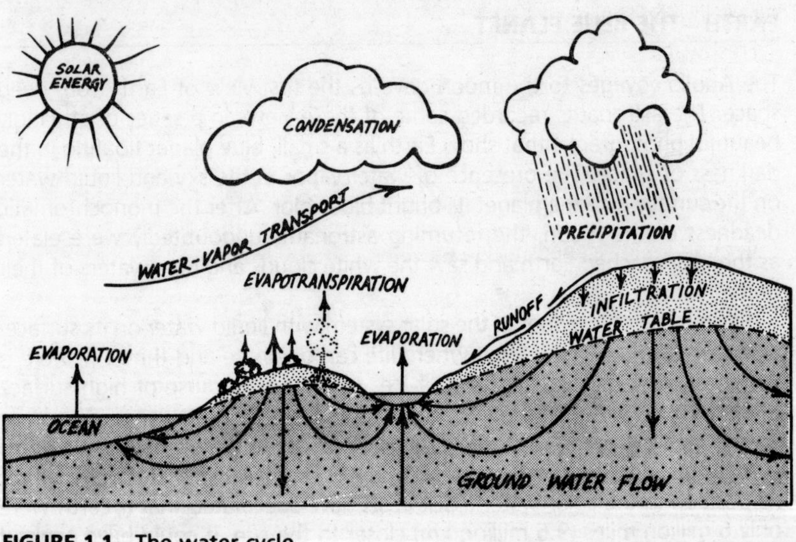

FIGURE 1.1 The water cycle.

TABLE 1.1 Estimated annual throughput to water cycle (source: R. L. Nace, U.S. Geological Survey Circular 536, 1967).

Water Item	Volume (km³)	Percentage of Total Water
Evaporation[a]		
From world ocean	350,000	0.026
From land areas	70,000	0.005
Total	420,000	0.031
Precipitation		
On world ocean	320,000	0.024
On land areas	100,000	0.007
Total	420,000	0.031
Runoff to oceans from rivers and icecaps	38,000	0.003
Groundwater outflow to oceans[b]	1,600	0.0001
Total	39,600	0.0031

[a]Evaporation (420,000 km³) is a measure of total water passing annually through cycle.
[b]Arbitrarily set equal to about 5% of surface runoff.

TABLE 1.2 Estimated total world water supply (source: R. L. Nace, U.S. Geological Survey Circular 536, 1967).

Water Item	Volume (km³)	Percentage of Total Water
Water in land areas		
Freshwater lakes	125,000	0.009
Saline lakes and inland seas	104,000	0.008
Rivers (average instantaneous volume)	1,250	0.0001
Soil moisture (above water table)	67,000	0.005
Ground water to depth of 4,000 m	8,350,000	0.61
Ice caps and glaciers	29,200,000	2.14
Total in land area (rounded)	37,800,000	2.8
Atmosphere	13,000	0.001
World ocean	1,320,000,000	97.3
Total all items (rounded)	1,360,000,000	100

Note: Figures are approximations and should not be taken as precise values.

cycle each year. Only about 5 of every 100,000 gallons of the total supply are in motion at any one time. The turnover rate for an individual particle of water in the atmosphere probably averages about 8–10 days, and it has been estimated that if all of the water in the atmosphere were to fall as rain on any one day, only about one inch of water would be measured at the earth's surface.

As shown in Table 1.2, most of the earth's water is stored in the ocean, all as salt water. Nearly all of the fresh water is frozen in the polar ice sheets. A measure of the water locked up in polar ice can be grasped when we consider that the Antarctic ice cap, if it melted uniformly, would yield enough fresh water to supply all the rivers of the world for about 750 years.

One could cite many other interesting facts about the water in storage and water in motion, but one fact stands out above all others: The continuous operation of the water cycle is absolutely essential for all life on the land. All land-based organisms lose water continuously to the environment, and most would die within a few days if they were unable to replace the lost water.

People were aware of the importance of the water cycle long before it was clearly understood. Some of our earliest ceremonial structures and rituals were aimed at recognizing and recording the changing seasons, which in turn signaled changes in warmth and in weather. More than 5,000 years ago the Egyptians measured fluctuations in the Nile and built dams and other structures to divert water to farmers' fields. At about the same time the Chinese were building various kinds of waterworks, and they may have been the first to use rain gages. The famous water laws of the Babylonian king, Hammurabi, sound strangely familiar to modern-day water users and regulators.

Although farmers and other water users possessed practical water knowledge for centuries, a thorough understanding of the water cycle came late in human history. The Greeks and their European successors, the scholastic philosophers of the Middle Ages, realized that springs and rivers must ultimately come from the ocean. But they seem not to have made the connection between rainfall and river flow or runoff. They thought rivers were fed from large springs or underground lakes, which in turn were somehow connected directly with the ocean. However, they were unable to figure out how the ocean water could lose its salt and how it could travel high into the mountains to feed river headwaters.

The birth of experimental science, during and after the time of the Renaissance in Europe, rekindled interest in natural phenomena. Leonardo da Vinci's notes revealed an apparent understanding of the water cycle as early as the fifteenth century. His knowledge was not widespread, however, and it was not until the latter part of the seventeenth century that the water cycle was firmly established in the scientific literature. Edmund Halley, the

English astronomer for whom Halley's Comet is named, showed that evaporation from the Mediterranean Sea was sufficient to account for water in the rivers emptying into the sea. In France, natural scientists Pierre Perrault and Edme Mariotte showed that rainfall in the Paris basin was more than enough to account for the runoff in the river Seine. Publication of these results during the last quarter of the seventeenth century, about 300 years ago, firmly established the water cycle as the central concept for the modern science of hydrology.

As the following chapters will demonstrate, hydrology involves a step-by-step study of the water cycle, including *evaporation, condensation, precipitation, infiltration, evapotranspiration, ground water,* and *runoff.* Most hydrologists specialize in one or more of these subjects, but the fields are so interconnected that one must have some understanding of them all to succeed as a hydrologist. Only recently have universities begun to offer programs in scientific hydrology. Many present-day hydrologists came from related fields such as meteorology, civil engineering, geology, and soil science.

Water in Motion

Although several factors contribute to water movement, solar energy is the main power source driving the water cycle. While the earth's gravitational field adds an important increment of energy, the sun is primarily responsible for the global circulation of water from sea to land and back to sea.

Oceans provide the main storage reservoir for water. While discussion of the water cycle could begin at any point, a logical starting point is the departure of a drop of water from the sea, through the cycle, and back to the sea. As it begins its travels, the drop of water leaves the ocean surface through *evaporation,* passing upward into the atmosphere as invisible water vapor. This process requires a large amount of heat energy, provided by incoming radiation from the sun.

Solar radiation provides practically all of the heat energy at the earth's surface; because oceans occupy about 70% of the earth's surface, they receive the majority of the sun's rays. In fact, an estimated 80% of incident heat energy from the sun goes to heat the surface water and hence to cause evaporation from the sea. Some heat does pass upward from the deep interior of the earth. However, the amount of geothermal heat is minuscule when compared with solar energy. The total amount of geothermal heat radiated at the earth's surface will evaporate about 1 millimeter of water per year.

According to the *conservation of energy* principle, energy can neither be created nor destroyed. What happened, then, to the energy that caused the drop of water to evaporate? It passed into the air with the water vapor, becoming the *latent heat of vaporization.* As it rises from the sea, the water vapor will be picked up in the global wind stream that blows continuously around the earth. At some point, the vapor-laden air may rise to a cooler region of the atmosphere where it can no longer retain its load of invisible water vapor. The vapor then gives up its latent heat of vaporization and reverts to a liquid state through *condensation*. If the temperature is low enough, it will freeze and become an ice crystal. Tiny particles called condensation nuclei facilitate the process, providing a surface on which the liquid water may form. These ever-present nuclei may consist of tiny salt crystals (borne aloft by wind during the evaporation process at sea), smoke, or dust particles from the land.

The *evaporation–condensation* process is an important factor in energy transfer at the earth's surface. Solar radiation passing through the air doesn't heat it very much. Most of the heat we feel on a warm summer day is either from sunshine falling on our bodies or reflected from the ground or other surfaces. Water vapor holds most of the heat in the lower atmosphere. The latent heat of vaporization, released to the atmosphere during cloud formation (condensation), is an important environmental factor in determining weather and climate. This kind of energy transfer can be powerful enough to drive hurricanes and other violent storms.

Even though our tiny drop of water is now visible, it would take a powerful microscope to see it—if it were alone. It is not, however; the drop is only one of billions joined to form a cloud. Droplets in clouds are so small that even though they have mass and are affected by gravity, the lightest breeze can keep them aloft indefinitely. To form rain or snow and fall as *precipitation*, many of these tiny drops must join to form drops or ice crystals large enough and heavy enough to fall from the cloud.

Before falling, the water droplet was affected mostly by thermal forces resulting from solar radiation. Once rain or snow begins to fall, the water enters the grip of the earth's gravity, and except for a possible short circuit back to the atmosphere through evaporation, gravitational forces will largely govern it until its final return to the sea.

The little particle of water has several avenues to follow once it reaches the ground. It could land on the leaf of a tree (a process called *interception*) and evaporate again to the atmosphere. It could drop on dry soil and go immediately into the ground in the process of *infiltration*. It could fall on a rock surface where it might begin to flow downhill toward a stream, initiating the process of *runoff*.

Water that falls on a running stream or on the land and immediately into a stream generally has a short journey back to the sea. Water that infiltrates the soil below the land surface, however, could go several ways. A plant could absorb it, carrying it upward. There the plant tissue could incorporate it, or the water could pass through the plant and out through the leaves, returning to the atmosphere in the process of *transpiration*. Water falling on a vegetated area may be lost to the atmosphere through transpiration or through evaporation from the soil surface. Since it is nearly impossible to separate these losses into two components, they are usually combined in one process called *evapotranspiration.* Water escaping evaporation from the soil or transpiration from growing vegetation may eventually pass down through the unsaturated soil zone to the *water table,* which is the top of the underground water body. Below the water table, water completely saturates the pores of soil and rock. The little drop of water would now have joined the component of the water cycle known as *groundwater.*

You may wonder how groundwater flows back into the ocean if it comes from rain on the middle of a continent far from the sea. You have probably had a picnic beside a creek that is flowing many days or weeks after the last rain storm. If no lake or reservoir is upstream, where does all that water come from? Its source is the only available water reservoir around—the groundwater reservoir. Water seeping out of the ground into the stream keeps most streams flowing between storms.

In any case, our little drop of water is now safely launched in a running stream and is finally on its way back to the sea. When it reaches its home in the sea, the circle will at last be closed; the water cycle will be complete.

2

The Structure and Properties of Water

What is water? Water is clouds, fog, dew, rain; water is frost, snow, hail, a glacier, an iceberg; water is a bubbling spring, a creek, a river, the ocean; water is part of every living thing; water is the stuff of life itself.

Almost everyone knows that water is a chemical compound made up of hydrogen and oxygen. Many people even refer to it in conversation as "H_2O" rather than using the word "water;" it may be the only chemical symbol they recognize. As the only naturally occurring liquid most people ever see, water is thought of as a common, ordinary substance, necessary but not at all unusual. Far from being ordinary, however, water is one of the most *extraordinary* substances on the earth. Life depends not only on the presence of water but also on some of its unique properties—its large heat capacity, its high surface tension, its lower density in the solid state, and its capacity as a universal solvent.

Understanding water's many unusual properties requires some knowledge of its internal structure. To say that water is made up of atoms of hydrogen and oxygen is only the beginning. The arrangement of these atoms sets water apart from other liquids.

MOLECULAR STRUCTURE OF WATER

The following account of water's internal structure is a brief, simplified discussion of a very complex subject. While the broad outline given here is believed to be essentially correct, scientists still do not fully understand the structure of liquid water. This need not deter us, however, from seeking a better understanding of water's unique properties through at least a rudimentary knowledge of its atomic and molecular structure.

Atoms consist of subatomic particles arranged the same general way in all atoms. Each contains a relatively heavy nucleus around which one or more electrons move. Hydrogen, the simplest of all atoms, has one outer electron; oxygen has eight. These atoms join to form a water molecule, sharing electron spaces in oxygen's outer rim and assuming an asymmetric shape (see Figure 2.1). Due to its shape and construction, one side of the molecule has a negative charge and the other a positive charge. The apparent positive and negative poles make this a polar molecule, with the positive end of one molecule attracting the negative end of another. This force of attraction is called a *hydrogen bond.*

The Hydrogen Bond

The hydrogen bond determines most of water's unique properties. It explains, for example, why water "wets" clean glass but will not stick to a glass surface coated with grease. The surface of glass has oxygen atoms with unpaired electrons, and water molecules form hydrogen bonds that hold a film of water to glass with a force much stronger than gravity. Because grease lacks these oxygen atoms with free electrons, water won't stick to greasy surfaces. Sticking to solid surfaces, or *adhesion,* will be discussed in more detail when we consider the phenomenon of capillarity.

Hydrogen bonding's strength in attracting individual molecules gives water a certain structural integrity. It literally "sticks" to itself—it is *cohesive*—thus imbuing water with a high surface tension. In a body of liquid water the positive end of one water molecule attracts (and bonds to) the negative end of another. At moderate temperatures these bonds continually form and break, resulting in a fluid character for the whole mass. The rate at which bonds form and break is a function of the energy in the system and depends on temperature. The higher the temperature, the greater the fluidity, as bonds form and break with increasing frequency. Warm water flows more

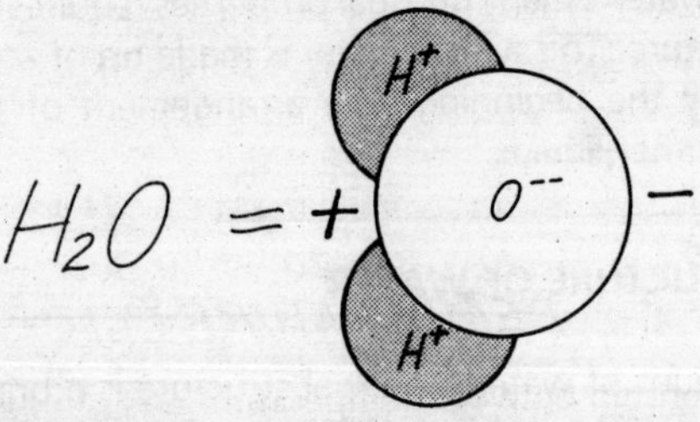

FIGURE 2.1 Diagram of water, a polar molecule.

readily than cold water, and at the freezing point water stops flowing altogether and crystallizes into ice.

Although we usually aren't aware of it, water expands as it heats and shrinks as it cools. This means that the density decreases with rising temperature and increases as the temperature goes down. Most other substances perform the same way. For most materials, however, as the temperature drops the density increases and volume decreases continually to the freezing point. Water is peculiar in that its maximum density occurs at several degrees above its freezing point (Figure 2.2). As the temperature in the liquid drops,

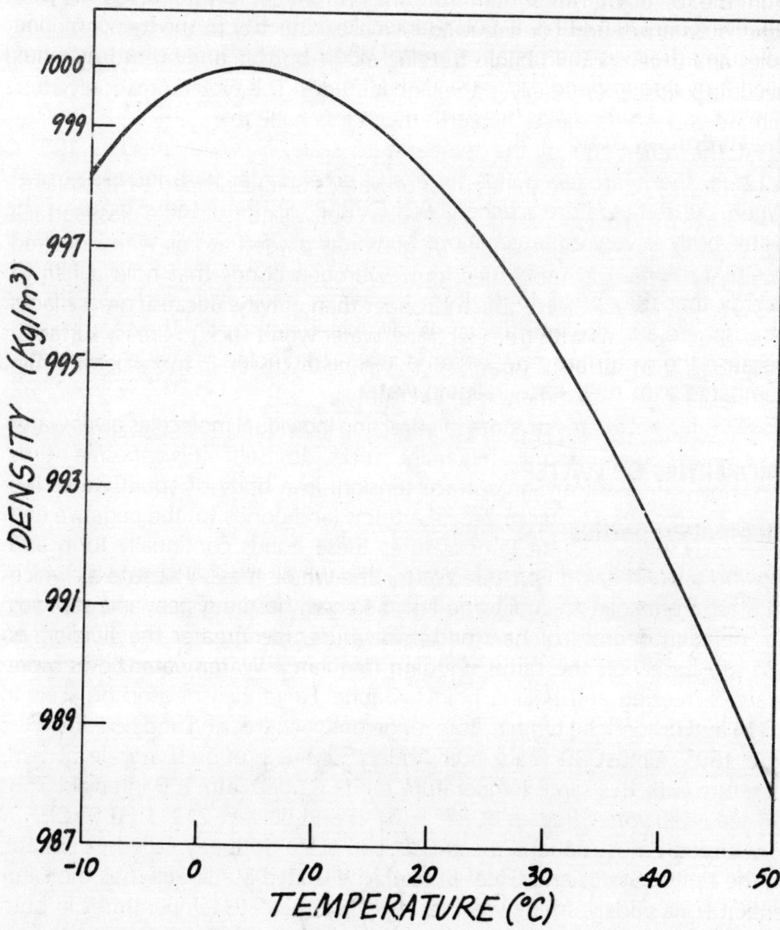

FIGURE 2.2 **Relationship between temperature and density of water at atmospheric pressure.**

molecular activity slows. The liquid volume shrinks as density increases and molecules move closer together. When the temperature of water reaches 4° C (39° F), hydrogen bonding becomes stronger than the liquid's tendency to shrink. Molecules are progressively arranged along lines of hydrogen bonds, resulting in gaps between the lines. This causes a decrease in density and an overall expansion in volume. As the temperature falls below 4° C (39° F), the water continues expanding until at 0° C (32° F) it freezes and forms the open, crystalline structure of ice.

Water volume increases about 9% when it changes from liquid to solid; this explains why ice floats in water and why bodies of open water freeze from the top down rather than from the bottom up. If water acted like most other substances and continued to increase in density to the freezing point, lakes and oceans would begin freezing at the bottom and eventually would become solid ice, especially in the high latitudes. This would drastically affect climate and would make the earth much less habitable.

At the other end of the temperature scale, as water reaches 100° C (212° F), the hydrogen bonds form and break faster with increased heat. When the temperature reaches 100° C (212° F), the internal state of the water body is very agitated, more bonds break than form, and the water begins to boil (at sea level). Large numbers of molecules receive so much energy that they break loose from the water body at its surface with the atmosphere and rise into the air as invisible water vapor. They are invisible because the water has now become a gas, the molecules widely separated compared with their state in liquid water.

PROPERTIES OF WATER

Thermal Properties

Boiling and Freezing Points. Water, because of its easily observed changes of state from solid to liquid and liquid to gas, has long been the standard for defining degrees of heat and cold. All temperature scales in common use are based on the same standard, the range in temperature between water's freezing and boiling points. Gabriel Fahrenheit devised his scale in 1714 and divided the temperature range between freezing and boiling points into 180°. Almost 30 years later Anders Celsius proposed a scale of temperature with this same temperature range divided into 100 intervals. Thus (at sea level) water freezes at 32° F (0° C) and boils at 212° F (100° C).

Although most nations use the Celsius scale (formerly called centigrade in the United States and Great Britain), the United States still uses the Fahrenheit scale widely. Many weather reports now state temperatures in both degrees Fahrenheit and Celsius, however; eventually, the Celsius scale may become the standard.

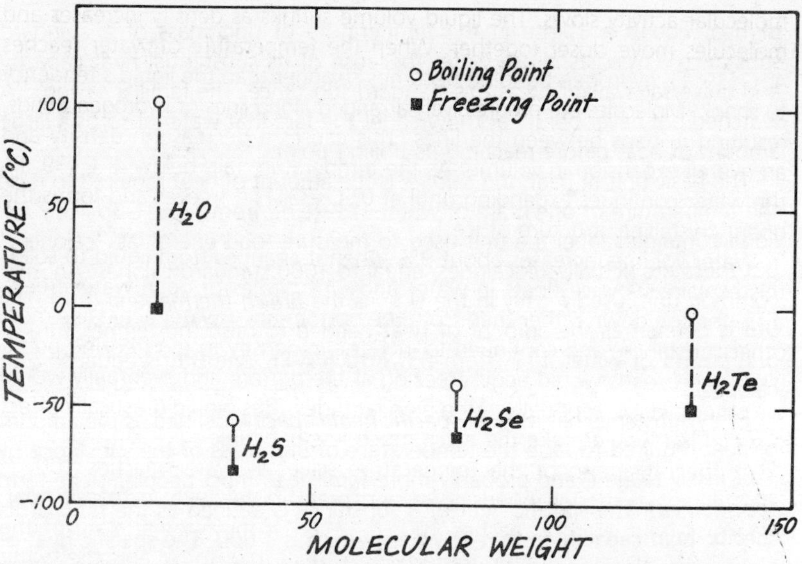

FIGURE 2.3 **Temperature range in the liquid phase for some hydrogen compounds of similar molecular structure.**

To see how much water differs from most other substances, we can compare its boiling and freezing points with those of some hydrogen compounds with similar molecular structure. Three such compounds are hydrogen telluride (H_2Te), hydrogen selenide (H_2Se), and hydrogen sulfide (H_2S).

A general principle that beginning chemistry students learn is that the higher the molecular weight of a substance, the higher its boiling point. For most substances in the liquid state this is true because intermolecular forces (called van der Waals forces) become stronger as the molecular weight increases. This is demonstrated for H_2S, H_2Se, and H_2Te in Figure 2.3, which also compares the three compounds with H_2O.

If water were an ordinary compound whose molecules were subject to the relatively weak van der Waals forces, its boiling and freezing points would fall somewhere below those of H_2S. Strong hydrogen bonding between the water molecules prevents that, however. Because of that bonding, water occurs in all three states (solid, liquid, and gaseous) at prevailing temperatures on the earth's surface. In a glass of water containing ice cubes, water vapor from the air will condense on the outside of the glass, demonstrating this quality.

Heat Capacity. Heat is a form of energy that is not measured directly but rather by how it changes the temperature in a substance or process. A unit of heat equals a 1° rise in the temperature of unit mass of a standard substance. Physicists chose water, because of its substantial heat capacity and universal availability, as this standard substance. The familiar saying, "A watched pot never boils," acknowledges the fact that water absorbs a large amount of heat before reaching its boiling point.

The basic unit of heat, the *calorie,* is the amount of heat required to raise the temperature of one gram of water 1° C (e.g., from 14.5° C to 15.5° C). Food containers label the unit used to measure food energy as "calories," but these are actually kilocalories, equal to 1000 standard calories. Another common unit, used mostly in the U.S., is the *British thermal unit* (Btu). The Btu is defined as the amount of heat needed to raise the temperature of one pound of water 1° F (from 63° F to 64° F); one Btu is equal to 252 calories.

Every substance has its own *specific heat capacity,* defined as the amount of heat required to raise the temperature of unit mass of the substance by 1°. Closely related, and probably more familiar to most people, is the term *specific heat.* The specific heat of a substance is defined as the ratio of its specific heat capacity to that of water, which is 1.000. The specific heat of a substance thus numerically equals its specific heat capacity, and since it is a ratio and therefore a pure number, a specific heat value will be the same in all systems of units. Table 2.1 lists specific heats for some common materials.

It should be noted in passing that the term "heat capacity" does not refer to how much heat a substance can hold, as in the case of a container that holds only a certain volume (e.g., a 1-gal. jug, 100-ml flask, etc.). Heat can

TABLE 2.1 Specific heat of some common materials.

Substance	Specific Heat	Temperature (in °C)
Water	1.00	14
Ice	0.50	−2
Glass	0.12	25
Mercury	0.03	25
Copper	0.09	25
Lead	0.03	25
Silver	0.06	25
Iron	0.11	25
Sandstone	0.26	100
Aluminum	0.22	25
Carbon (Diamond)	0.12	25
Carbon (Graphite)	0.17	25
Alcohol (Ethyl)	0.53	0

be added to a substance indefinitely, causing, of course, a corresponding rise in its temperature and perhaps eventually a change of state.

To see what this property of materials means in practical terms, consider aluminum. As Table 2.1 indicates, the specific heat of aluminum is about one-fifth that of water. This means that the same amount of heat will cause a five-fold increase in the temperature of aluminum as compared with water. For example, heat that causes a 10° increase in water's temperature will cause the temperature of aluminum to rise by about 50°. Anyone who has ever tried to drink hot coffee or tea from an aluminum cup has already discovered this fact. The cup also feels so hot because of aluminum's high heat *conductivity,* which is about 300 times that of water. The low heat capacity and high thermal conductivity make aluminum an ideal metal for cooking utensils.

Heat of Fusion and Heat of Vaporization. Water's high heat capacity is due in large part to its internal structure and the peculiarities of hydrogen bonding. As the temperature rises, adding more heat energy to water, the molecular bonding progressively deteriorates in what one author has called "structural melting." Closely associated with this are two more of water's unusual properties, the latent heat of fusion and the latent heat of vaporization.

By definition the *latent heat of fusion* is the amount of heat per unit mass required to change a substance at its melting point completely to a liquid at the same temperature. The *latent heat of vaporization* is the amount of heat per unit mass necessary to change a liquid at its boiling point completely to a gas at the same temperature. The word *latent* denotes a change of state without a change in temperature. For water at the melting point the latent heat of fusion is about 80 calories per gram (144 Btu/lb); at the boiling point the latent heat of vaporization is about 539 calories per gram (970 Btu/lb).

Ice water in a glass stays at a constant temperature as long as unmelted ice is still floating around in the liquid. Likewise when a teapot is boiling away on the stove, the water and the teapot remain at constant temperature as long as any water remains in the pot. If the water should boil away, however, the metal pot might turn red hot or even melt on the stove. As the ice was melting in the glass or the water was boiling in the pot, heat was being added to the system without changing the temperature (Figure 2.4). Energy was being absorbed, first as latent heat of fusion and then as latent heat of vaporization. Because these processes are reversible at ordinary temperatures on the earth's surface, they are probably two of the most important energy transformations in our environment.

While vaporization requires the addition of heat energy, this process occurs at temperatures well below 100° C (212° F). It requires merely enough internal energy in the liquid water body to break hydrogen bonds holding

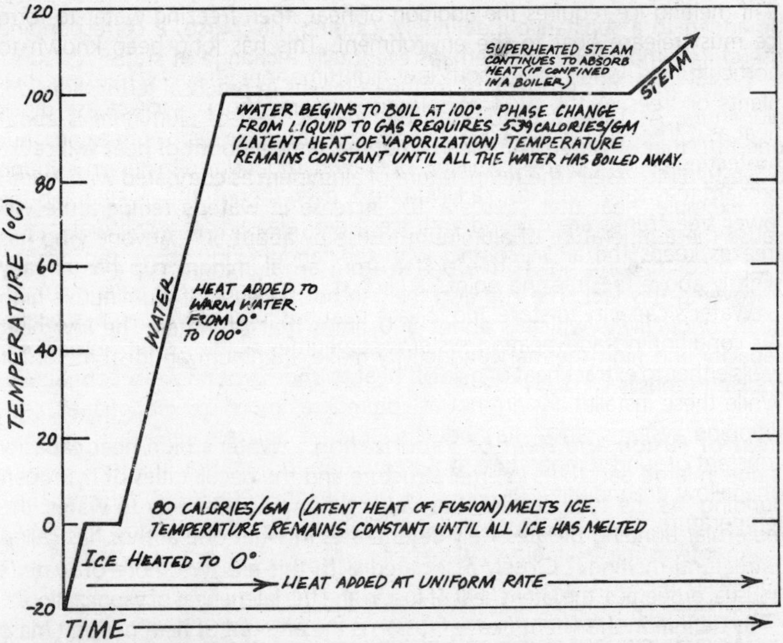

FIGURE 2.4 Energy relationships as water changes state from solid to liquid to gas.

a molecule in the surface and to release it as a gas molecule into the atmosphere. It is important to remember that regardless of the water body temperature, each molecule of water vapor takes with it some heat, the latent heat of vaporization—according to the conservation of energy principle, heat energy added to the water to cause vaporization cannot be destroyed. Thus, if added heat does not increase the temperature of the water during evaporation, it must still be with the vapor in the atmosphere.

As noted previously, fusion and vaporization are reversible; these processes of energy transformation largely run the earth's weather and climate machine. Evaporation from the sea transforms solar energy (which had heated surface water) into latent heat of vaporization. The water vapor, carried aloft by winds, may eventually condense to form clouds. During condensation, the vapor releases its latent heat energy to the atmosphere. Where much evaporation and condensation occur, as in the tropics, an enormous amount of energy is released to the air. This may cause the air to become very warm; when excessive heat energy is released to the air above tropical oceans, violent storms such as hurricanes sometimes develop.

If melting ice requires the addition of heat, then freezing water to form ice must release heat to the environment. This has long been known to horticulturists, who worry about low nighttime temperatures freezing their plants or trees. A tub of water placed in a greenhouse will release much heat as it freezes and may keep the air temperature just above freezing, thus protecting delicate plants. Similarly, operating irrigation sprinklers in a young orchard may save trees from a killing frost, even though the ground and the lower tree trunks are covered with ice. Heat released from the water as it freezes keeps the air in contact with the tender buds above the ground slightly above the freezing point.

Water's capacity to store and release heat has long been used practically. Air conditioning and heating systems sometimes employ heat pumps and wells either to extract heat or give off heat to underground water formations. While these installations are not as common as more conventional air conditioning systems, they may be very efficient and cost effective, given a constant ground temperature. It is interesting to note that the temperature of shallow groundwater usually approximates the region's mean annual air temperature.

An interesting example of water's capacity to absorb and store heat occurred in the groundwater body beneath a city in the northwest United States following World War II. Several buildings constructed after the war were equipped with air-conditioning systems that used groundwater from wells to help cool the air. Water passing through the air conditioners was discharged to the sewer and did not reenter the ground. As the systems operated only during the summer, the wells drew water from the groundwater body in the summer only. The river flowing through the city also recharged the underground formations in the summer, replacing water withdrawn by the wells. Since the river's summer temperature is quite warm, the continual replacement of well water actually raised the groundwater temperature several degrees over a period of years. Little water was pumped other than that used for summer air-conditioning. With its large heat capacity, the groundwater retained summer heat from the river indefinitely. A front of warm water from the river traveled slowly, year after year, through the gravels underlying the city, moving toward the wells. In time warm river water replaced much of the original cool groundwater under the city, and the air-conditioning systems became less efficient.

The same kind of operation, only in reverse, has surfaced in recent years as a way to store heat from solar energy collectors. Hot water would be pumped from the collectors into underground formations, which would store the heat for use at a later time—for example, at night, when solar collectors on the roof would be cold.

If you observe processes taking place around you, both in nature and in human technology, you will see countless examples of how these unique thermal properties of water benefit us. You will also see how our intervention in natural systems sometimes produces unexpected, sometimes unwelcome, side effects.

Viscosity

While the word *viscosity* may not be the most familiar one in your vocabulary, the property is likely to be familiar in your experience with liquids. When pouring honey or molasses out of a pitcher, you know that it flows more easily when it is warm (low viscosity) than when it is cold (high viscosity). Viscosity affects the way motor oil lubricates your car engine, and auto manufacturers usually recommend specific viscosities for various driving conditions. Low-viscosity oils are recommended for cold weather and higher viscosities for hot weather.

One can think of viscosity as the internal friction of a fluid. That friction requires that a force be applied to cause one layer of a fluid to slide past another. When molasses is poured out of a pitcher, the force is gravity. For motor oil, the force is supplied by the oil pump on the engine. As the temperature increases, viscosity increases for gases and decreases for liquids. For example, water is about three times more viscous at 20° C (68° F) than it is at 80° C (176° F); hence it would flow more readily at 80° C than at 20° C.

Although you may have been aware that molasses or motor oil exhibit viscosity, you probably didn't realize that water does, too. For a change of 1° F (0.5° C) in water temperature, the viscosity will change about 1.5%, rising as the temperature drops and falling as the temperature increases. Consequently, over the temperature range usually encountered in nature, an increase of 1° F (0.5° C) theoretically will boost the flow rate about 1.5% because of decreased viscosity.

Some years ago a groundwater infiltration system was built to collect water from gravels beneath a river to supply a city in Washington. The system operates using gravity, with the natural head of water in the river providing the energy to cause water to pass from the river bed, through the underlying gravels, and into perforated pipes connected to a deep caisson or shaft dug in the river bank. Water flows from the river, through the gravel, and into the caisson continuously all year. It is then pumped into the city water mains. The only variables are river level and temperature. An accurate flow meter measures the yield where it discharges into the city mains. During different seasons the yield varied considerably, even when the river level was essentially the same. In winter, with a river temperature of 40° F (4° C), the collector yielded 4.86 million gallons per day (mgd) (18,397,044 L/d) and in the summer, when the river was at 54° F (12° C), the yield was 6.18 mgd (23,393,772

L/d). That impressive difference of 1,320,000 gallons (4,996,728 L) in 24-hour production of water was due entirely to a change in river temperature.

Sometimes water's properties can be put to practical use in ways that seem strange and far from ordinary experience. Working in cooperation with the New York Fire Department, a large chemical corporation invented a formulation that reduces the natural viscosity of water and makes it flow faster at a given pressure in a fire hose. Firefighters can now get the same amount of water from a 1¾-inch (4.4 cm) hose as they did before from a 2½-inch (6.4 cm) hose. The smaller hose is easier to carry, bends around corners more readily, and generally offers a much better way to fight fires.

Compressibility

Have you ever heard of compressed water? Probably not. How about compressed air? Very likely you have, especially if you drive an automobile. Occasionally we must reinflate auto tires using an air compressor, the kind found at gasoline service stations. What is compressed air? It is simply ordinary air that a compressor has squeezed into a smaller volume by confining it in a container under pressure. Gases such as air are relatively easy to compress because their molecules are far apart; pushing them together does not demand an inordinate amount of energy. The molecules of liquids, on the other hand, are already close together. Common experience teaches that liquids are incompressible. Otherwise, how could a hydraulic press transfer and magnify force through applying pressure to a liquid confined in the fluid system of the press, or a car's braking system, or the many hydraulic systems in a modern aircraft? While the hydraulic system transfers force without any *apparent* reduction in volume, in fact a very small amount of compression occurs in the liquid. For water at ordinary temperatures the compressibility factor is about 0.0000034, meaning that a hydrostatic pressure of 1 pound per square inch (6.89 kilopascals) would reduce unit volume by this amount, about 3 one-millionths of the original volume.

For all practical purposes water can be considered incompressible. On a global scale this property of water is significant. For example, it has been estimated that if water really were incompressible, the oceans would stand more than 100 feet (30 m) higher and nearly 2 million square miles (5,000,000 km^2) of low-lying coastal lands would be flooded. Of less consequence but still important is the effect of compressibility on the storage capacities of deep underground water formations.

Surface Tension

Have you ever filled your coffee cup or water glass right up to the rim so the liquid rises above the edge as though a thin skin or membrane was holding it there? Or have you ever dropped a few drops of water onto a hot grill and then watched it dance and bounce around in little spherical drops until it finally evaporated and disappeared? Probably everyone has watched water dripping from a faucet and seen the drops assume a spherical shape as they fell through the air. These are examples of *surface tension,* which results from the cohesive forces within a body of water.

When water comes into contact with air, the surface molecules are subject to very uneven forces. Down in the body of the liquid molecules attract all the surrounding molecules. But at the air-water surface a molecule is only partly surrounded by other water molecules. Consequently, it is attracted only to the body of the liquid. This tends to draw the surface molecules inward and makes the liquid act as though an invisible membrane enclosed it.

"Floating" a steel needle demonstrates the surface's considerable strength (Figure 2.5). If the needle were turned on its end and put in the water, it would penetrate the surface and sink to the bottom. Not enough surface molecules would be in contact with the small end of the needle to support it. When it is gently placed on the surface lengthwise, though, many molecules tied together with powerful hydrogen bonds bear its weight. It really does float on the water. Figure 2.5 shows how the needle depresses the water surface, and the arrows show the direction of forces in the surface caused by the needle's weight. It appears that a thin membrane is being stretched or put in tension; hence, the name "surface tension."

Surface tension characterizes all liquids, differing greatly in magnitude for different liquids. Because of hydrogen bonding the surface tension of water is two or three times higher than that of most common liquids. (Liquid metals

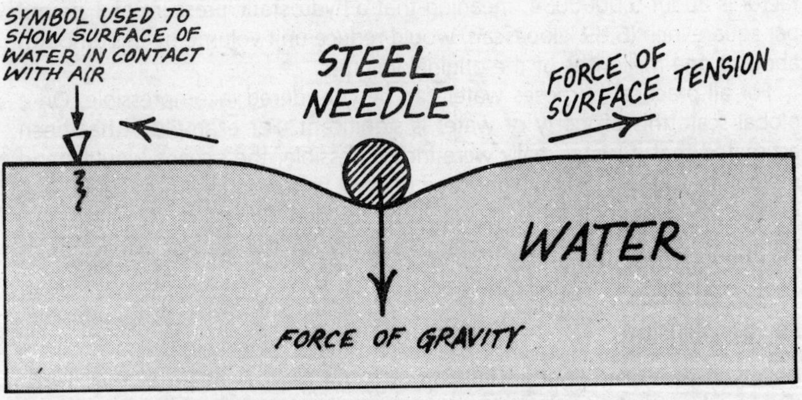

FIGURE 2.5 **Steel needle supported by surface tension of water.**

are one exception, and the surface tension of mercury is more than six times that of water.) The surface tension of water decreases with increasing temperature, and it may also be affected by chemical additives. For example, adding a little detergent drastically lowers surface tension by making the hydrogen bonds less effective.

Capillarity

Capillarity results from a combination of surface tension (cohesion) in the water and its tendency to wet solid surfaces (adhesion). As hydrogen bonds form between the water and the solid, the water tends to "climb" up the wall. At the same time, cohesive forces in the body of the liquid are trying to draw the surface molecules into a minimum configuration. A curve in the water's surface forms that is most noticeable in a small-diameter glass tube, a so-called capillary tube (see Figure 2.6). If the tube were placed in liquid mercury, which does not wet glass, the surface tension would draw the liquid surface down instead of up, and the meniscus (curved surface) would be convex instead of concave, as it is with water. The height that water rises in small tubes depends on the tube diameter and on the water temperature. For example, at ordinary temperatures, water will rise a little over 3 centi-

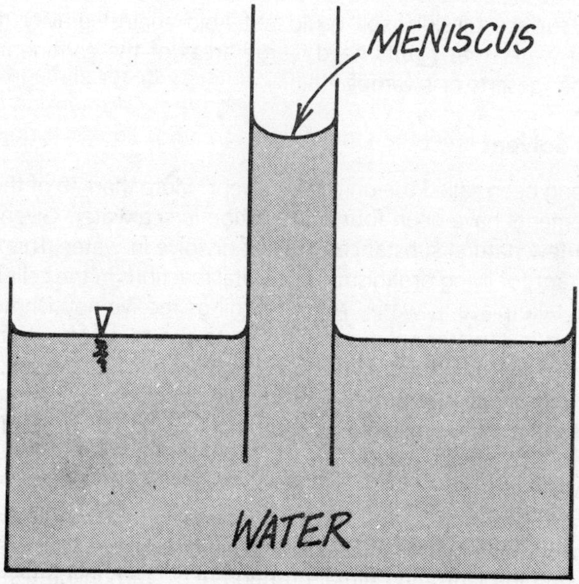

FIGURE 2.6 Capillary rise of water in a small glass tube.

meters (1.2 in.) in a glass tube of 1 millimeter (0.04 in.) diameter. In a ⅛-millimeter (0.005-in.)-diameter tube, the height of rise will be more than 25 centimeters (9.8 in.). In very small tubes the capillary rise may be several feet above a free water surface. This is especially important to the circulation of water in soil and of blood in the body.

An interesting example of capillary forces working in soil is loose sand in which the tiny pores between sand grains act as tubes of small diameter. Think now of being at the beach. Dry sand can be piled up at a slope of about 30°, its angle of repose. Put more sand on the pile and it simply slides downhill; the slope remains stable at about 30°. If you dig down to where the sand is moist, however, you can make a trench in which the sand is perpendicular to the surface. What's the difference between dry and moist sand? If you could examine the texture of the moist sand under a magnifying glass you would see the difference. Thousands of tiny menisci hold the sand grains together, each one exerting the strong forces of hydrogen bonds to hold the capillary water to the grains and hence binding the grains. If more water is added—if a wave washes up and saturates the sand trench—the vertical face of sand collapses and slumps back to an angle of repose. The menisci and capillary forces holding the sand grains together were destroyed when the mass was saturated, and gravity pulled the sand down.

Surface tension and the related phenomenon of capillarity have enormous implications for the success of life on the land. If water had a much smaller (or weaker) surface tension, soil could not hold against gravity the large quantities of water that plants need. Large areas of the continents would become either deserts or swamps.

Water as a Solvent

Water has long been called the universal solvent. More than 70 of the known chemical elements have been found in solution in sea water. Given enough time, only a few natural substances will not dissolve in water. This is particularly important for living organisms, since vital reactions in the cells of plants and animals take place in water. All living things receive nourishment from food delivered in a water solution, whether to the roots of a plant or the bloodstream of our own bodies. In the inorganic world, conversely, water is one of the most corrosive substances known. In terms of geologic time, water can dissolve away limestone to form mammoth underground caverns, such as Carlsbad Caverns in New Mexico. Eventually it can destroy even our most massive works.

In general, water acts as a solvent in two ways. First, water dissolves a substance by forming hydrogen bonds with its molecules. Many of the compounds needed in plant and animal nutrition (e.g., carbohydrates) are held in solution this way. Second, water molecules surround individual ions of the substance going into solution. *Ions* are the electrically charged parts of molecules that become separate entities during the solution process. The ions may consist of single atoms or of several atoms in electrically charged units. An example is sodium chloride (common table salt), wherein the sodium (Na^+) has a positive charge and the chlorine (Cl^-) has a negative charge.

When sodium chloride (NaCl) goes into solution it separates into two ions, which are surrounded by water molecules arranged so that the negative side (oxygen) of some water molecules are around the Na^+ ion and the positive side (hydrogen) of other molecules are around the Cl^- ion. The sodium and chloride ions are thus kept apart in separate little "cages" and are prevented from recombining to form the compound sodium chloride. Only after the water has evaporated will the Na^+ and the Cl^- combine again as NaCl.

Substances like NaCl that dissociate into ions are called electrolytes, and water has a substantial capacity for dissolving electrolytes. One gallon (3.8 L) of water (8.3 lb or 3.8 kg) will dissolve several times its own weight of some common electrolytes. For example, 1 gallon (3.8 L) of water will dissolve as much as 70 pounds (32 kg) of the common fertilizer, ammonium nitrate (NH_4NO_3). In solution the NH_4NO_3 dissociates or comes apart as ammonia (NH_4^+) ions and nitrate (NO_3^-) ions.

Many gases and solids dissolve in water without increasing its volume. A pound of ammonium nitrate dissolved in a gallon of water will increase the water's weight by one pound but will cause essentially no increase in its volume.

Electrical Properties

Water's electrical properties are closely related to its solvent properties. Although hydrogen and oxygen atoms in a water molecule have an unbalanced electrical charge and in that sense could be called ions, the molecules are attached to each other by hydrogen bonds and are not free to move around separately as ions in an electrolyte are. For that reason water alone cannot carry an electrical current, and very pure water is a good insulator—it is said to have a large dielectric constant, which allows it to neutralize the attraction between oppositely charged ions, as in the dissociation and dissolving of sodium chloride. On the other hand, electrolyte ions in a solution, although they are shielded from coming in contact with each other by the surrounding water molecules, are free to move toward an electrical pole placed in the water. In a sodium chloride solution the negative chloride ions (Cl^-) move toward the positive pole and the positive sodium ions (Na^+) move toward the negative pole, thus causing an electrical current to flow through the solution. The more ions in the solution, the more conductive the water becomes. For example, sea water is many times more conductive than ordinary river water. By measuring electrical conductivity we can even estimate the total quantity of electrolyte contained in a water solution.

SUMMARY

Water could exist on the earth without life, but there would be no life without water. Life began and evolved on the earth because of the presence of liquid water in large quantities. When organisms were ready to move from the sea to the land, about 400 million years ago, they found a friendly environment created long before by the operation of the water cycle. Fresh water flowed over the land in streams, and porous soil held water ready to nourish primitive plant roots. The continued, steady operation of the water cycle over time has allowed the slow processes of organic evolution to create the tremendous diversity of living forms now inhabiting the earth. Water's crucial properties and their role in this evolutionary sequence will become evident in succeeding chapters.

3

Water in the Air: Evaporation and Condensation

. . . there ariseth a little cloud out of the sea. . . .

(I Kings, 18:44)

The preceding quotation was part of a report to Elijah by his servant, who had been sent to the top of Mt. Carmel to observe the weather. Elijah was concerned about a coming storm; shortly "it came to pass . . . that the heaven was black with clouds and wind, and there was a great rain."

Have you ever said, "It surely looks like it's going to rain"? Why did you say that? What did you see or feel that made you think rain was coming? Probably it was just water in the air, either invisible water vapor or visible clouds, but water nevertheless. It has been said that when we look out at the weather, we see some form of water—clouds, rain, snow, and so on. Viewing weather that way makes it easier to relate the weather and the water cycle. If weather is mainly "water in the air," a study of the water cycle can begin there as well as anywhere, using evaporation as the first step.

EVAPORATION

The results of evaporation—changing visible water into invisible vapor—must have been clear to people even in ancient times. A thoughtful observer watching water disappear from a hot rock surface or from a boiling pot would soon realize that it must be disappearing somehow into the air. At least 2,500 years ago the Greeks understood the concept of evaporation as the process that raised water up out of the sea into the atmosphere. An understanding of the mechanics of evaporation developed relatively recently.

The Process of Evaporation

It is now well known that vaporization, the change of state from liquid to gas, involves molecules escaping from a water surface into the overlying air. As all molecules in the liquid are in constant motion, it follows that only especially energetic molecules will be able to break all bonds with their neighbors and move up into the air. Remember that latent heat of vaporization (about 539 calories per gram at 100° C) determines a change from liquid to vapor. The thermal energy of latent heat is merely transformed kinetic energy from fast-moving molecules. That this removes energy (and hence heat) from the liquid is evident, considering that evaporation tends to reduce the temperature of the liquid phase, as you have no doubt noticed when you stepped out of the shower. Evaporation rapidly cools the thin film of water still on your body.

Once the water molecule becomes a gas molecule it behaves differently than it did in the body of the liquid. Each molecule in a gas is more isolated and much further from neighboring molecules than those in a liquid. In relatively humid air on a typical summer day the vapor molecules above a lake surface will be more than 40 times as far from their neighboring molecules as those in the water below. Hence hydrogen bonding properties, which are so important in liquid water, will be weak or absent altogether in the vapor phase.

As the average spacing between molecules increases during evaporation, water undergoes a huge expansion. At ordinary temperatures 1 gram of liquid water occupies approximately 1 milliliter. At 25° C (77° F) 1 gram of saturated water vapor occupies about 42,000 milliliters, an expansion of about 42,000 times. No wonder the vapor molecules are far apart and not affected by hydrogen bonding. It is also interesting to note that water vapor is only about 60% as heavy as other atmospheric gases, and it tends to rise in the air above a water surface.

Except for its high thermal energy content, water vapor shares many of the properties of other common gases. It can compress or expand, and its molecules exert a pressure of their own called "partial pressure." From experiments conducted in 1801, English chemist John Dalton developed the law of partial pressures: "In a gas mixture the molecules of gas of each kind exert the same pressure as they would if present alone; and the total pressure is the sum of the partial pressures exerted by the different gases in the mixture." This means that in the air above a water surface only the partial pressure of water vapor is important, and effects of the other atmospheric gases may be ignored. The vapor pressure of water at different temperatures and pressures has been determined experimentally, and tables of these values are readily available. As with other gases, the volume of a given weight of water vapor will vary with the prevailing temperature and pressure, but only the ordinary atmospheric temperatures and pressures prevailing at the earth's surface are of interest to hydrologists.

A given volume of air can hold just so much water vapor at any given temperature and pressure; and this property is referred to as *humidity.* Saturated air has a relative humidity of 100%. A humidity reading of less than 100% indicates that the air could hold more moisture. In general, the higher the temperature, the more moisture the air can hold. For example, air with a relative humidity of 100% at 40° F (4.4° C) would have a relative humidity of only 50% if the temperature were raised to 60° F (15.5° C). That explains why early morning fog evaporates and disappears as the rising sun begins to warm the air.

Theoretically, at 100% humidity the number of molecules leaving the water surface would balance those returning from the air, and net evaporation would be zero. This might happen in the laboratory, but in nature (where air is free to move) evaporation will most likely continue as long as energy is being added to the water, the evaporation rate varying according to the environment. Heat energy is essential for evaporation; its source is generally solar radiation. The air's relative humidity and motion across the water surface will also influence the rate of evaporation. Near the equator, where the sea and winds are warm, evaporation lifts much water into the air. Tropical climates are typically humid, as in the East Indies, the Amazon basin, and in central Africa. Near the poles, where the cold winds can't hold much moisture and the weak sunshine gives off little heat, little evaporation takes place. Arctic climates are typically dry: Alaska's North Slope, despite the spongy muskeg in summer, has an arid climate.

Effect of Water Quality. Water quality has a small but measurable effect on evaporation rates. A salt content of 1% will slow the rate of evaporation by about 1%. Since the ocean generally has a salt content of slightly more than 3%, sea water evaporation rates lag freshwater rates by about the same 3%.

Effect of Water Depth. As explained in Chapter 2, water has the capacity to absorb heat energy. The larger the water body, the more energy it can store. A deep lake, then, will contain more heat in storage than a shallow lake, and that stored heat will influence how fast water evaporates from the two lakes. To understand why this is so, consider what happens as seasonal temperature changes heat or cool the surface water.

In a cool climate during winter an entire shallow lake may be near 39° F (4° C). When the surface freezes, the temperature will vary from 32° F (0° C) just under the ice to a constant 39° F (4° C) a few feet down. As spring turns into summer the air becomes warmer and warmer, and so does the surface of the lake. As the ice melts and the surface water warms from 32° F (0° C), the surface water becomes heavier, sinks to the bottom, and is replaced by cooler, lighter water rising from below. This turnover continues until all the water is at 39° F (4° C), the temperature of maximum density for water. As the temperature rises above 39° F (4° C), the water becomes lighter and remains at the lake surface. As the season advances and the air temperature rises, surface conduction will slowly heat the deeper water.

In the autumn the whole process is reversed: Surface water cools, becomes heavier, and sinks to the bottom. This continues until all the water is at 39° F (4° C). After that, surface water becomes lighter as it cools from 39° F (4° C) to 32° F (0° C), and it remains at the surface, where it may eventually freeze during the coldest part of winter. In a warmer climate water goes through the same kind of seasonal turnover, but at a higher temperature range and with little or no surface freezing.

Consider a deep lake as winter wanes and warm spring winds begin to blow over its surface. Assuming that the main body of water in the lake has reached thermal equilibrium during the winter, the rise in surface water temperatures will be slow and will affect only the upper levels. Below about 200 feet (61 m), the water remains constantly at its minimum temperature, a fact of no little importance to aquatic life in a river below a large dam that releases water from the bottom of the reservoir. In Lake Mead, the reservoir behind Hoover Dam, a one-year study showed that the surface temperature averaged 68° F (20° C) and water below 200 feet (61 m) averaged 52° F (11° C). Raising a large lake's temperature requires a huge amount of heat; consequently, water temperature changes will lag seasonal changes in the air temperature. Temperatures in the shallow lake, on the other hand, correspond much more closely to the air temperatures.

Seasonal evaporation rates from deep and shallow lakes reflect their different temperature regimes. Figure 3.1 compares annual evaporation for Lake Superior (deep) and Lake Heffner (shallow).

Lake Superior, the westernmost of the Great Lakes, is a large, deep lake with a very large volume of water. Lake Heffner is a shallow water-storage reservoir for Oklahoma City. (We will have more to say about Lake Heffner shortly, but now we are interested only in comparing evaporation records from Superior and Heffner.) While climates at the two localities are not exactly alike they are not too dissimilar. Both are in the central part of the continent and are affected by weather systems passing across North America. Air temperatures at Lake Heffner are consistently higher, although both localities

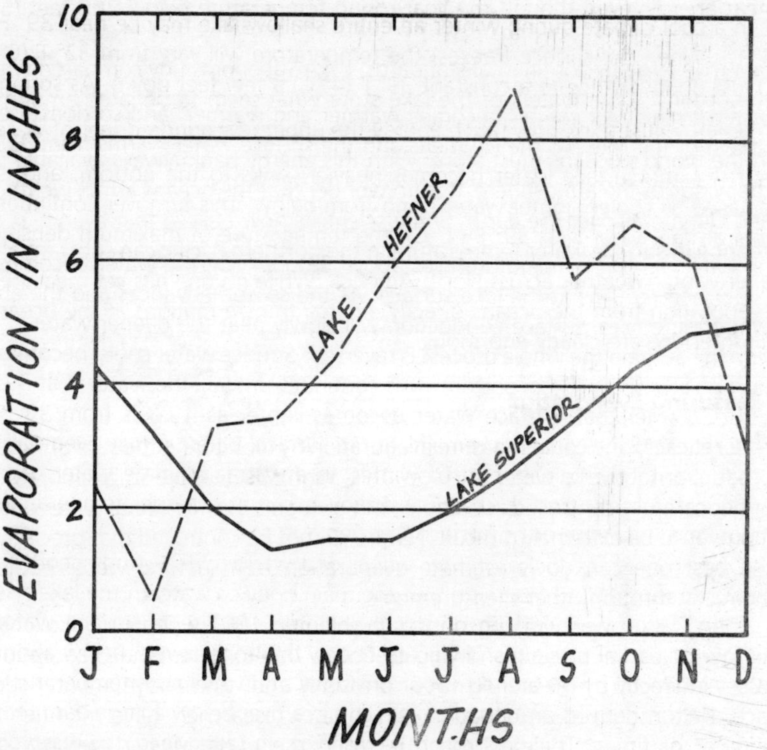

FIGURE 3.1 Annual evaporation from Lake Superior and Lake Heffner.

have definite variations from winter to summer. Annual evaporation at Lake Heffner is about 57 inches (1450 mm); at Lake Superior, about 36 inches (910 mm).

It is in water volumes—and hence in heat storage capacity—that the two lakes differ so drastically. The total storage capacity of Lake Heffner, with a surface area of 4 square miles (10 km²), is a small fraction of 1 cubic mile of water. Lake Superior, on the other hand, with a surface area of 31,820 square miles (82,407 km²), contains a few *thousand* cubic miles of water. In summer Lake Heffner's temperature lags the air temperature by about 15 days, and its limited capacity to store heat from the sun means a high summer evaporation rate. The winter rate is relatively low because it depends on daily solar radiation, since the body of the lake stores little heat.

Lake Superior's water temperature lags far behind the air temperature, and as Figure 3.1 shows, evaporation is actually higher in the cool months than in midsummer. With a maximum depth of 1290 feet (393 m), the lake has a large volume of deep water, which in turn stores a large amount of heat energy, even though the year-round temperature below 200 feet (61 m) is probably close to 39° F (4° C).

While large by continental standards, Lake Superior is tiny compared with the oceans. If the waters of the lake store what seem to be large amounts of energy, just think how much energy the upper few hundred feet of water in the world's oceans must store. With this energy bank always available to the winds of the atmosphere, it is no wonder that the oceans affect weather and climate all over the earth. Some scientists now believe that even a 1° C change in surface water temperature in the northern Pacific can substantially alter winter weather across North America. While no one has yet measured evaporation from the oceans directly, energy transformations at the ocean-air interface are clearly enormous.

Measuring Evaporation

Theoretically, one could measure evaporation by monitoring the rate of vapor production above a water surface. This might be feasible in a laboratory under carefully controlled conditions, but outdoors over a natural water body it proves to be extremely difficult. Nature cannot be controlled that precisely; the best one can do is estimate evaporation from natural water bodies indirectly through other measurements.

John Dalton was working on this problem in 1801 when he discovered his law of partial pressures. Trying to find a way to determine evaporation losses indirectly by measuring vapor pressures and wind above a water surface, Dalton derived an equation, which later researchers have modified a number of times. Their approach is called the *mass-transfer method* for estimating evaporation.

A second method, called the *energy-budget method,* draws on the conservation of energy principle. It states that the total incoming energy must equal the outgoing energy (including latent heat of vaporization) plus any increase in the water body's internal energy.

Both methods are theoretically sound and can be applied using available measuring instruments. Applying either method to a real water body, however, is both time consuming and costly. A more practical method, developed in the United States, uses a small metal pan of standard construction along with some very sophisticated mass-transfer and energy-budget experiments to calibrate the pan measurements.

National Weather Service Class A Pan. Various pans are used to measure evaporation throughout the world. In the United States a standard pan has been used for many years. It is called the National Weather Service Class A pan (see Figure 3.2). This pan, which is both economical and easy to use, is

FIGURE 3.2 National Weather Service Class A evaporation pan with anemometer, hook gage in stilling well, and thermometer (courtesy Belfort Instrument Company).

48 inches (122 cm) in diameter and 10 inches (25 cm) deep and is constructed of galvanized sheet metal.

When the station is in operation, the water depth is kept between 7 and 8 inches (18–20 cm) in the pan. Daily measurements of the water level are made with a hook gage in a stilling well (see Figure 3.3). Evaporation is computed as the difference in the daily water levels. Another method, using only the stilling well, involves adding a measured amount of water each day to bring the level to a fixed point in the stilling well. Either method requires correcting the computed evaporation for any precipitation that occurred during the period.

Although government bureaus often equip evaporation pans with an anemometer to record total wind passage (as shown in Figure 3.2), a recording thermometer, and an instrument to record solar radiation, the only instrument required is a rain gage. Since the aim is to measure total water loss, obviously any rainfall must be deducted. The key element is probably the pan's specific position in the local environment. A pan placed on a paved driveway or parking lot will give a disproportionately high reading, while one in the shade of a tree will give a similarly low reading. The preferred location is out in the open on level ground with grass or other low vegetation upwind.

A.

B.

FIGURE 3.3 Hook gage and stilling well. The stilling well provides an undisturbed water surface around the hook gage, which is graduated in either English or metric units (courtesy Belfort Instrument Company).

The pan is placed on a wooden base (see Figure 3.2) to allow air to circulate underneath. A fence should protect the site from animals, although birds are nearly impossible to keep out, because most observers don't want to cover the pan. Even putting wire mesh over the pan changes the evaporation rate. If a few birds splash some water out, that's a source of error most observers are willing to accept.

As Figure 3.2 indicates, because the pan is small, its energy relationships only vaguely resemble those found in a large lake or reservoir. Scientists and engineers who have used pan measurements to estimate lake or reservoir evaporation have thus applied a coefficient or multiplying factor to the pan measurements to reflect more closely evaporation from the larger water body. A proper value for this coefficient has generated a great deal of discussion over the years. The Lake Heffner experiment, undertaken shortly after the end of the Second World War, settled the issue to the satisfaction of most scientists.

The Lake Heffner Experiment. The original problem was an attempt to estimate the evaporative water loss from Lake Mead, the reservoir behind Hoover Dam in the Southwest. Lake Mead was then the largest artificial body of water in the world. It is the key reservoir controlling distribution in the lower Colorado River for Arizona, southern California, parts of Nevada, and Mexico. These are arid regions; all water supplies are precious. Lake

Mead proved to be too large, however, and the problems of measuring water losses were too complex. A smaller, simpler lake was needed; Lake Heffner was chosen.

Lake Heffner, located almost in the center of Oklahoma, is the reservoir supplying Oklahoma City. A small lake, with a surface area of about 2300 acres (931 ha), it is so situated that there is very little underground seepage. Practically all water lost from the lake is either measured discharge to the city water system or evaporation losses. This is important for checking the accuracy of experiments for estimating evaporation. A number of instruments for doing mass-transfer and energy-budget studies were situated in, over, and around the lake to measure incoming solar radiation, back radiation from the lake, air and water temperatures, wind speed, rainfall, and so on. Several government agencies jointly conducted the studies over a 15-month period from April 1950 to August 1951.

The Lake Heffner results were published as U.S. Geological Survey Professional Paper 269 (1954). The experiment showed that both the mass-transfer and energy-budget methods held much promise as efficient means of estimating evaporation from lakes or reservoirs. Since then, follow-up studies done at Lake Mead and throughout the United States have confirmed the original findings at Lake Heffner. Of particular interest were the conclusions regarding evaporation pans.

Although several types of pans were calibrated during the experiments, most interest centered on the National Weather Service Class A pan. A chief advantage of this pan, particularly in the United States, is its long history of use compared to all the other kinds of pans. Data are published annually for about 500 stations; Figure 3.4 shows the pattern of evaporation across the country, as measured in Class A pans. Most estimates of reservoir evaporation in the United States have depended on Class A pan measurements. Furthermore, the American Society of Civil Engineers—an organization whose members perform many of the independent evaporation measurements for reservoir studies—recommends the pan.

Although the Class A pan coefficient was found to vary (seasonally) at Lake Heffner, the recommended figure for annual evaporation that most engineers use is 0.70. In other words, observed Class A pan evaporation (over a 12-month period) multiplied by 0.70 should yield a reasonable estimate of annual evaporation from a nearby lake or reservoir. Experienced engineers believe that estimates based on a pan coefficient of 0.70 ordinarily would not be in error by more than 15% from the true reservoir evaporation.

Estimating Evaporation from Weather Data

Beginning with John Dalton's studies in 1801, a number of attempts have been made to devise a formula for estimating evaporation using either

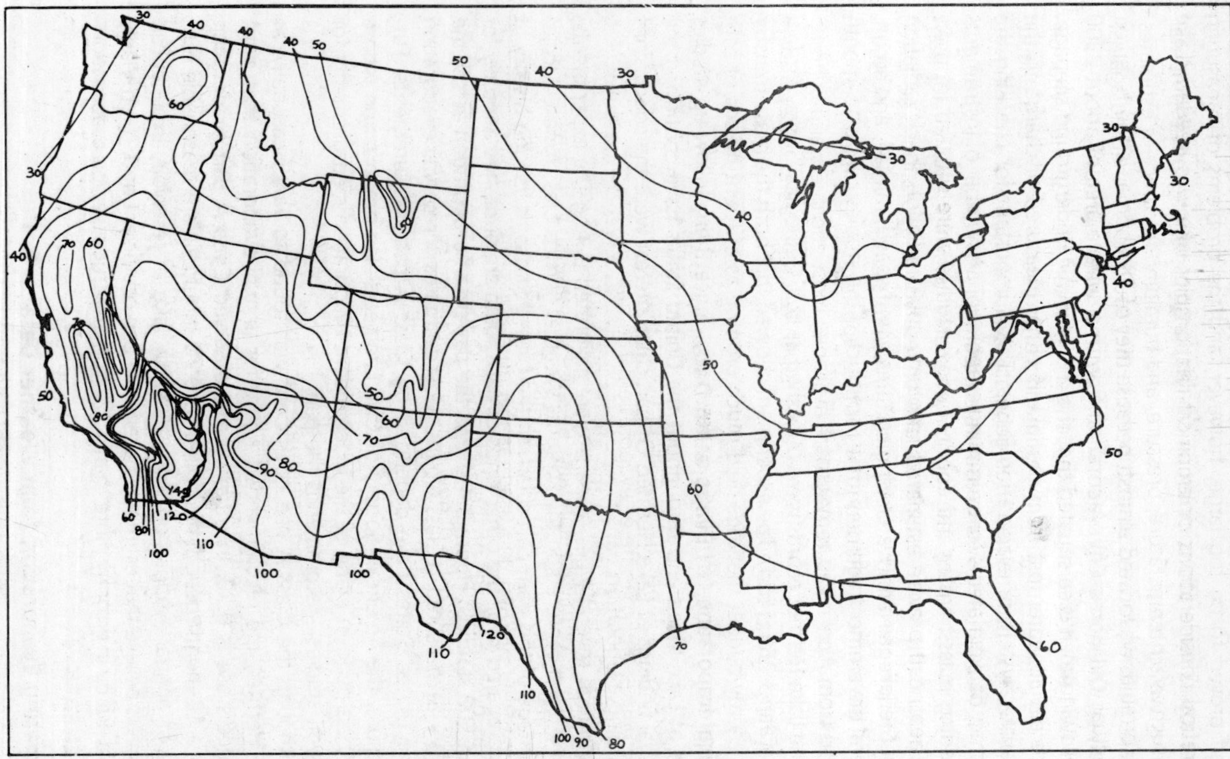

FIGURE 3.4 Mean annual pan evaporation (in inches) (source: National Weather Service).

theoretical considerations or empirical relationships among environmental data such as air temperature, pressure, and humidity. Accuracy of estimates from formulas was always suspect because they could not be checked against the actual (but unknown) evaporation from real water surfaces.

Following the research at Lake Heffner, however, the National Weather Service developed a method for estimating evaporation in a Class A pan at a given locality. The method uses generally available weather data or data that may be estimated with reasonable accuracy. As with actual measurements from Class A pans, the formula estimates (with the appropriate pan coefficient) can provide estimated lake or reservoir evaporation. The Lake Heffner report given in U.S. Geological Survey Professional Paper 269 (1954) explains the method. Although the formula approach is valuable where no actual evaporation measurements are available, most engineers and hydrologists still prefer the record from a Class A pan when planning or designing large water-storage projects.

Practical Importance of Evaporation Data

Pan observations may be made daily or at less frequent intervals, depending on the season and the rate of evaporation. The data are recorded in inches or millimeters of water lost to the atmosphere during a specified period—daily, weekly, monthly, or annually. Because of seasonal variations in evaporation rates, especially from large bodies of water, annual totals probably are most often quoted and used (as in Figure 3.4). The annual figures are useful as overall indexes of the wetness or dryness of a region or locality. They also give a rough idea of how efficient a projected water storage project might be. For example, the U.S. Geological Survey has estimated that when water is released at a rate of 450 cubic feet per second (12.7 m^3/s) from Twin Lake Reservoir to the Colorado irrigation canal network, an average of about 16% of the water is lost in a 12-day transportation period. Evaporation, transpiration from vegetation along the canal banks, and bank and channel storage produce the losses. Depending on flow conditions, high air temperatures, and distance the water must travel, losses may run as high as 28% of the original reservoir release. Viewed this way, evaporation losses are clearly an important consideration in determining feasibility of water resources projects in arid regions.

CONDENSATION

Condensation—the change from gas to liquid—is the reverse of evaporation. When a gas, in this case water vapor, is cooled sufficiently, condensation returns it to the liquid state. The *heat of condensation* released to the air exactly equals the *heat of vaporization* that the water vapor acquired during evaporation. Most of the time as we watch clouds form in the sky the rise in air temperature due to the heat of condensation goes unnoticed. On the rare occasions when large amounts of heat are released rapidly, however, the air becomes very unstable and violent storms may result. Thunderstorms, tornadoes, and hurricanes are all chiefly the result of rapid, large-scale condensation of water from warm, moist air. Together, *evaporation-condensation* is one of the most important energy-conversion processes at the surface of the earth. It provides much of the power to run the earth's weather machine and bring a never-ending supply of fresh water from the oceans to the land.

The Process of Condensation

Wherever water is in contact with air, some molecules are always moving out of the water into the air and some returning from the air to the liquid. Theoretically, when water vapor saturates air under stable conditions, the number of molecules leaving the liquid balances the number returning. In reality, most of the time the balance is tipped one way or the other through either evaporation (more molecules leaving water) or condensation (more molecules returning to water).

Although it is physically correct to define condensation this way, that is not how most people think of condensation. Most are more familiar with formation of clouds or fog in the atmosphere or of water droplets on the outside of a cold bottle of beer on a warm day. In all these cases the process is the same. Invisible water vapor releases its latent heat and becomes liquid water again. For this to happen in nature, the moisture-laden air must be cooled until it reaches saturation (100% relative humidity), and a surface must be present on which the water can condense. The temperature at which saturation occurs and condensation begins is called the *dew point*. The name describes a common occurrence on chilly mornings when air near the ground is cooled by radiation from the earth toward a clear sky and dew forms on the grass as the air is cooled below its dew point. These same conditions may also produce a shallow ground fog.

Condensation Nuclei. When dew forms, it is easy enough to see the surfaces on which water condenses: a blade of grass, leaves of plants, or sometimes the metal or glass surfaces of an automobile parked outside overnight. When the ground fog forms, though, on what kind of surfaces in the air was water able to condense? They are so small we can't see them, yet they always seem to be there. Meteorologists solved this mystery long ago by collecting samples of air on very fine filters and examining them under powerful microscopes. They found that what looks like clear air is actually peppered with extremely small particles. These tiny particles, which provide the surface on which water forms, are called *condensation nuclei*. They contain all sorts of things—salt crystals from the sea, dust from the ground or from volcanic eruptions, smoke particles, and even tiny liquid droplets discharged from industrial waste stacks or cooling towers. The last type of nuclei may be particularly effective (because condensation is more likely to form on a tiny particle of water than on a solid surface), and when atmospheric conditions are right, one can often see a plume of fog forming downwind from a large oil refinery or power station.

Formation of Clouds and Fog. Condensation on an extensive surface such as a blade of grass or a glass bottle causes easily visible drops to form. Condensation on nuclei in the air results in drops that, if alone, would be as invisible as the bare nuclei. They are not alone, however. When a parcel of moist air is cooled to the dew point, condensation takes place suddenly, forming a cloud with billions of microscopic droplets. All those droplets together tend to scatter and diffuse the light, making the cloud visible. Fog is the same thing, just closer to the ground.

Both fog and clouds are visible signs of the atmosphere cooling. Air always has a supply of condensation nuclei and almost always enough moisture to form clouds if the temperature drops low enough. When the temperature is very low, as in the high atmosphere or in the polar regions, clouds form from tiny ice crystals rather than water droplets. Clouds usually form due to cooling by vertical ascent of air masses, and one classification of clouds is based on the kind of air movement that caused the cooling. Since clouds are responsible for precipitation, a broad classification of storm types is closely related to cloud types; this will be discussed more fully in the next chapter.

Clouds and fog form when the temperature drops to the dew point, but what happens if the temperature rises above the dew point or continually changes as the air moves? When air temperature rises above the dew point, the cloud droplets evaporate and disappear; the morning fog "burns off" as the sun rises higher in the sky. You have seen the morning fog disappear. Have you ever lain on the grass and idly watched cumulus clouds float by in a summer sky? When you focus on one cloud awhile, you see the edges growing or contracting. Time-lapse photography shows clearly what is happening: Clouds are continually forming and dissipating. Meteorologists say that the life of a typical small cumulus cloud is probably only about 5–15 minutes. Similarly, in a cloud that seems to remain over a mountain front for hours, a typical cloud droplet lasts only about 10 minutes. On the other hand, some of the layer-type clouds that accompany weather fronts may last for several days. Understanding how and why clouds form (or dissipate), indicates a great deal about what is going on up in the atmosphere.

Measurement of Condensation

Condensation is one phase of the water cycle that is seldom measured. An intermediate step between evaporation and precipitation, it is usually thought of simply as water in transit—clouds in the sky waiting to rain. In extremely dry regions, however, even the moisture of fog or dew is important. Israeli scientists have actually measured approximate quantities of condensation in the form of dew by observing how much dew forms on a specially prepared surface exposed to the atmosphere at night. In the United States very light precipitation such as dew is sometimes measured by a "dew balance" that weighs and records the amount of dew deposited on a special nylon sieve. Measurements have also resulted from observing and comparing the differences in growth rates of plants exposed to dew versus plants protected from dew. Plants exposed to dew at night (along the coastal plain of Israel) grew about twice as much as those protected from dew.

Along the dry coast of South West Africa/Namibia, rainfall is low (about 1 inch—25 mm—or less per year). The cold Benguela current in the South Atlantic, together with warm desert air, produces much fog during some seasons. Some plants and animals in the Namib desert depend almost entirely on fog and dew for their water supply. One of these is the rare, ancient *Welwitchia mirabilis*, which Charles Darwin called the "platypus of the plant kingdom." The welwitchia depicted in Figure 3.5 is thought to be more than

FIGURE 3.5 ***Welwitchia mirabilis*, growing in the western Namib Desert a few miles east of the coastal town of Swakopmund. This specimen, shown here in a fenced enclosure, is thought to be about 1500 years old.**

1500 years old; it is living evidence of the climate's stability along this dry coast.

APPLICATIONS

Knowledge of the evaporative process can be used to practical advantage in at least three ways:

1. Preventing loss of usable water through suppressing evaporation.
2. Disposing of unwanted water by promoting evaporation.
3. Using the energy transformation during evaporation (the latent heat of vaporization) to cool water heated by industrial processes.

Suppressing Evaporation to Reduce Water Loss

Evaporation is a natural process that occurs wherever liquid water and sufficient energy exist to cause a change of state from liquid to vapor. While we can never stop the process, it is sometimes possible to at least retard it, thus reducing loss of usable water. The idea is not new. More than 200 years ago, Benjamin Franklin experimented with oil films to suppress evaporation from a pond in England. In recent years scientists have experimented with chemicals such as hexadeconal (cetyl alcohol), which will form insoluble layers one molecule thick on the surface of a pond or reservoir. Less than 10 grams of this chemical will make a film covering about 43,000 square feet (4,000 m^2) of water surface. As you might suppose, factors such as wind and water currents complicate matters, but where the film can remain unbroken over the entire surface of a reservoir, it can reduce evaporation losses (theoretically) by as much as 70%. Actual field trials in Australia have shown 20%–40% water savings. As the cost and availability of water become more critical in arid regions of the earth, more effort will be made to conserve supplies; consequently, chemical films may be used more extensively for evaporation suppression.

Although chemical films have been found useful in reducing evaporation losses, this technique requires more or less continuous surveillance to ensure that the monomolecular layer on the water surface remains unbroken. Such surveillance increases the cost of the water saved, and in remote locations constantly monitoring a reservoir site may be too costly. This is especially true where a reservoir is filled only infrequently by flash floods in desert water courses. A typical region is South West Africa/Namibia, where water engineers have devised an ingenious structure called a sand-storage dam to save precious runoff water.

Sand-storage dams were first built by German engineers in the early days of this century when South West Africa was a German colony. They soon proved their effectiveness in saving runoff water that would otherwise have evaporated, and a number of them have been built during the past several decades. Construction progresses in stages; its purpose is to build a barrier across a dry water course where sand will be carried downstream and trapped during the next flash flood (see Figure 3.6).

In South West Africa/Namibia the climate is extremely dry, and except at its borders it has no perennial surface streams. A few large dams and reservoirs have been constructed on the principal drainage courses for large-scale water storage, mostly to supply towns and mines. These dams are of conventional design and the reservoirs are like any others in a desert climate. Evaporation losses are naturally high; but with a large store of water, engineers can cope with such losses. Small reservoirs have proportionately higher evaporation losses. Therefore if a farmer or village water department plans a conventional open-water reservoir on a small drainageway, evaporation may result in major water losses. If, on the other hand, the reservoir is in the intergranular pores of a saturated sand body, salvaging most of the runoff water trapped behind the dam is very likely. Once the water level in the sand drops about 2 feet (0.6 m) below the upper sand surface, evaporation virtually ceases. Total water volumes in sand-storage reservoirs are not large by ordinary reservoir standards, but in the desert any amount of water saved is a valuable asset. An additional benefit results when the reservoir lies over a permeable base, allowing some of the stored water to percolate downward to recharge the regional groundwater body.

Disposal of Wastewater by Evaporation

Where the climate is favorable, evaporation ponds can be an efficient means for disposing of unwanted wastewater. Keeping water in the pond shallow promotes a high evaporation rate. In shallow depths, incident solar energy will go into producing vaporization rather than stored in a large body of deep water. The ponds must be sealed to prevent the wastewater from percolating down into the underlying groundwater. Furthermore, enough ponds are necessary to accommodate sufficient storage capacity for the winter months when evaporation is at a minimum. The method works best in a warm, arid or semiarid climate, as illustrated by the example shown here from California's San Joaquin Valley.

Disposal of Oil-Field Wastewater. The southern San Joaquin Valley contains most of the petroleum reserves of California, and the daily production runs into hundreds of thousands of barrels. Because most oil occurs in marine sedimentary rocks, oil wells nearly always produce some salt water along with the oil. The oil and water are separated at the surface, and while the oil is shipped to the refinery, the water remains in the oil fields and must be

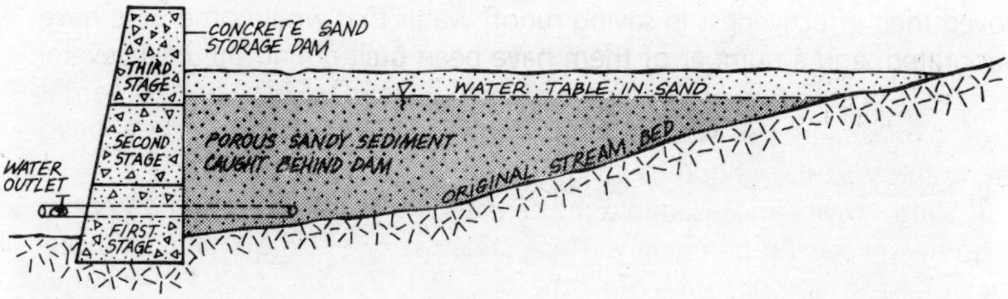

FIGURE 3.6 Sand storage dam across water course in South West Africa/ Namibia. Sketch shows how dam is raised in stages to trap sand carried downstream by flash floods. Recharge to underground storage occurs during annual floods. Stored water is drawn from outlet at bottom of sand behind dam (source: Water Research Commission, Republic of South Africa).

eliminated. Oil fields in the San Joaquin Valley are situated in the midst of one of the richest irrigated U.S. agricultural regions, and salt water must be disposed of so that it does not contaminate surface water or groundwater used for irrigation. Only three possible methods for disposal exist: (1) transport the wastewater out of the area using pipelines; (2) inject the water into deep, salt water formations through wells; or (3) store the wastewater in shallow, sealed ponds and let it evaporate. Transporting water out of the area would be prohibitively expensive. Some water is injected back into the ground to aid in enhanced oil recovery from the deep formations, but this is also expensive. Wastewater in many of the oil fields is currently being disposed of by evaporation or, as described in Chapter 6, by evapotranspiration. A series of evaporation ponds near the town of Maricopa at the southwest margin of the valley (see map in Appendix I), operating for more than 20 years, has proven efficient and cost-effective in disposing of oil-field waters (see Figure 3.7). Evaporation here is about 90 inches (2286 mm) per year.

Disposal of Agricultural Wastewater. Evaporation ponds may also dispose of agricultural wastewater. In recent years, farmers in the southern San Joaquin Valley have been forced to install shallow subsurface drains to lower perched water tables (see Chapter 7) and remove saline water from waterlogged fields. The drain water is collected in a system of ditches and canals that empty into large sumps (of several hundred acres), which act as evaporation ponds for final disposal of the saline water. Eventually the California Department of Water Resources plans to construct a master drain from the southern valley to the Sacramento River where it enters San Francisco Bay. Until such a drain is built, however, the local evaporation ponds are an effective means for disposing of the unwanted saline water.

FIGURE 3.7 Evaporation ponds for disposal of the oil-field waste waters near Maricopa, California, looking northeast across oil fields with agricultural lands in the distance. Although this photo was taken in 1963, the ponds are still effectively diposing of oil-field waters (photo by California Dept. of Water Resources, Courtesy of Valley Waste Disposal Company).

Evaporative Cooling of Industrial Process Water

The cooling effects of evaporation were well known even in ancient times. People learned that porous clay pots, in which water seeps through the sides and evaporates, tended to keep water cooler than pots with glazed sides. In modern times the water bag, made from coarse fabric with a porous texture, has been widely used by travelers in the desert. Before the ice chest was invented, most people who drove through the deserts of the western United States had a water bag hanging somewhere on the outside of their cars (see Figure 3.8). In more remote regions of the earth, where ice is not readily available, the water bag is still a common sight in the desert.

Applying the same principle, but on a much larger scale, evaporative cooling is widely used in industry to cool water. Many industrial processes discharge large quantities of waste heat. Steel mills, power stations, oil refineries, natural-gas compressor stations, and so on all require cooling systems as part of the process. Where a nearby large river or the ocean can supply cold water, a once-through cooling system is feasible. Where large supplies of water are not available for cooling purposes, though, various closed systems may be used. Cooling towers, spray ponds, and similar

A.

B.

FIGURE 3.8 Thirsty geologist taking a pull on his water bag on a hot June day in the Nevada desert. Water in bag is kept much cooler than outside air temperature by evaporation as water seeps through coarse fabric of bag (photo courtesy of Robert T. Littleton).

techniques that use evaporation for cooling the water are common, especially in large electric power stations.

Figure 3.9 is an aerial photo of the Rancho Seco nuclear power station a few miles southeast of Sacramento, California (see map in Appendix I). The size of the cooling towers indicates the waste heat that must be absorbed by the environment at a nuclear power plant. Only about one-third of the thermal energy liberated during nuclear fission is converted to electricity; the rest must then be disposed of in the environment. Water for this plant comes from an irrigation canal that brings water from the American River, several miles to the north.

The cooling towers shown in the photo are the natural-draft type, which introduce hot water near the top, allowing it to trickle down over a system of baffles. The baffles separate the water stream into smaller parts and droplets to expose more hot water to the upward-moving air current. The towers have open latticework at the bottom and act much like chimneys in promoting a natural draft of air through the inside.

After water passing through the towers is cooled, it is collected in reservoirs (at the left of towers in the figure) for recirculation through the power plant. Water lost during the cooling process must be replaced continuously. In a natural-draft tower, evaporation losses are about 1% for each 10° F (4.7° C)

FIGURE 3.9 **Aerial view of Rancho Seco Nuclear Power Plant near Sacramento, California.**

drop in water temperature, and windage losses out the top are probably slightly less than 1% of the hot water introduced into the tower. These figures demonstrate why cooling towers, or similar evaporative systems, are popular where water is scarce. Towers come in all sizes, from the giant towers at a nuclear plant to the small towers at an isolated compressor station on a natural-gas pipeline.

4

Precipitation

"If the clouds be full of rain, they empty themselves upon the earth."

(Ecclesiastes, 11:3)

"Skies will be overcast tomorrow with a 90% chance of rain."

(TV weather forecast)

The two quotations above may prompt you to ask the question, "Have people really changed all that much in the last 5,000 years?" In respect to their interest in the weather, the answer would be "No." The modern sciences of hydrology and meteorology give us a better understanding of the water cycle than our ancestors had, but their interest in nature was surely just as keen as ours. Without the help of modern science they made observations of natural phenomena that sometimes came close to matching what we do today. Nowhere is this demonstrated better than in the measurement of precipitation.

A recent research report by two Chinese scientists (Ren and Liu) shows that the Chinese have observed seasonal temperatures and weather for about 3,000 years. It is reported that they also measured and recorded rainfall for at least part of that period. We know definitely that rainfall was measured on an organized basis in India as long ago as the fourth century B.C., and Palestinians used rain gages in the first century A.D. Europeans began to use rain gages late in the seventeenth century, and today precipitation is routinely measured in all parts of the world.

The hydrologist's interest in precipitation focuses on water that reaches the ground as rainfall and snowfall. While hydrologists are not directly concerned with what happens in the atmosphere prior to arrival of water on the land, they still need to understand these processes to interpret precipitation measurements.

THE PROCESS OF PRECIPITATION

Because precipitation results from moisture in the air, it is intimately associated with the general circulation of the atmosphere. The circulation begins in the region around the equator. Hot sun overhead causes warm, moisture-laden air to rise to high altitudes where it flows away from the equator toward the poles (Figure 4.1). Due to the earth's rotation, the air moves northeasterly in the Northern Hemisphere and southeasterly in the Southern Hemisphere. It cools as it moves away from the equatorial regions, and at about 30° north or south it begins to sink. Near the earth's surface part of the air continues on toward the poles and part returns toward the equator. These air currents form the familiar trade winds and prevailing westerlies. At the same time polar air is joining an opposite circulation where very cold, dry, heavy air sinks and flows along the surface toward the equator. When lobes of cold polar air move far enough south (or north) to confront the poleward-moving tropical air, stormy weather usually results. Local conditions may affect the general circulation to some extent, and since most of the land is in the Northern Hemisphere, weather patterns there are different

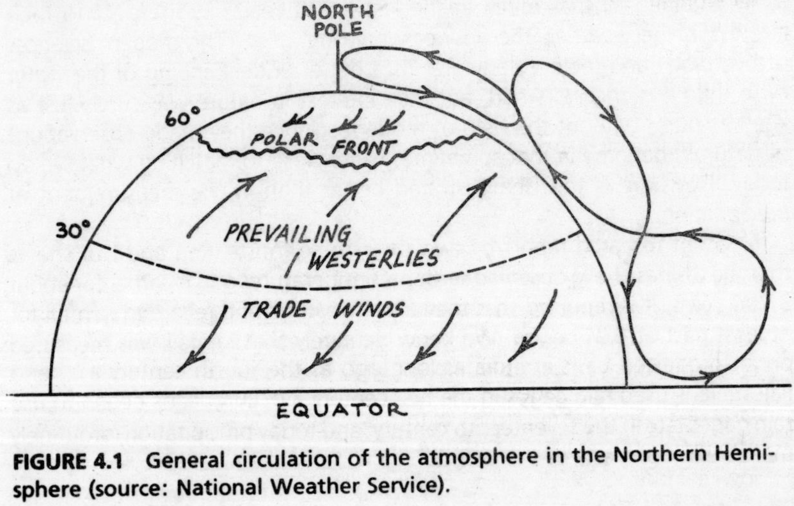

FIGURE 4.1 General circulation of the atmosphere in the Northern Hemisphere (source: National Weather Service).

from those in the Southern Hemisphere, where so much of the earth's surface is ocean.

The main source of precipitation, even over the middle of continents, is water vapor derived by evaporation at the ocean surface. As this moist maritime air passes over the land there are several ways in which it can be cooled below its dew point to form clouds. All of the ways involve raising the moist air to higher elevations. The air can be cooled by being lifted up the slope of a mountain range, or up and over a tongue of colder (and heavier) polar air, or straight up by heated air rising in the process of thermal convection. In all cases the end result is the same: The air cools below the dew point, moisture condenses on the ever-present condensation nuclei, and a cloud forms.

The visible moisture in a cloud is made up of tiny water droplets or ice crystals averaging about 0.01 millimeters (0.0004 in.) in diameter. In completely still air these droplets or crystals would eventually settle to the earth. Air is never completely still, however, and even the faintest breeze is enough to keep cloud droplets suspended. To produce precipitation, the droplets must move together to form large drops or crystals that are heavy enough to fall to earth. This can happen in two ways: by coalescence or by growth of ice crystals. In clouds with temperatures above freezing, individual droplets in the cloud collide and coalesce to form raindrops. In some cases ice crystals and supercooled water droplets coexist in clouds in which the temperature is below freezing. When this happens, the liquid drops tend to evaporate and recondense on the ice crystals, thus enlarging them to a size heavy enough to fall from the cloud. Most raindrops average about 1 millimeter (0.04 in.) in diameter, and about a million cloud droplets together make one raindrop. These processes require great activity in the cloud, and not every cloud will produce precipitation. Even the darkest, wettest cloud will not rain for long on its own. Meteorologists have estimated that if all of the moisture in the atmosphere were to fall suddenly, only about one inch (25 mm) of rain would fall over the earth's surface; however, records show that sometimes many inches of rain will fall in one hour. This is possible only if a continual inflow of moist air recharges the rain clouds.

Forms of Precipitation

Only water that falls to the earth is considered precipitation by hydrologists. This would exclude fog, dew, and frost, even though these nonprecipitating forms of condensed atmospheric moisture nourish many organisms. The common forms of precipitation are drizzle, rain, ice pellets, snow, and hail.

Drizzle. This is a fine mist with drops just a little larger than heavy fog, about 0.1–0.5 millimeter (0.004–0.02 in.). The intensity of precipitation is low, seldom resulting in more than 1 millimeter (0.04 in.) of water per hour.

Rain. This designation applies to all liquid precipitation heavier than drizzle. Rain drops average about 1 millimeter, but may range from 0.5 to 5 millimeters (0.02–0.2 in.) in diameter. Intensities of rainfall are reported (in the United States) as *light* when the rate is less than 0.10 inches (2.5 mm) per hour, *moderate* for rates between 0.11 and 0.30 inches (2.8–7.6 mm) per hour, and *heavy* when rainfall is more than 0.30 inches (7.6 mm) per hour.

Ice Pellets. Also called *sleet,* these occur when raindrops freeze as they fall through air where the temperature is below 0° C (32° F). Ice pellets are transparent spherical grains of ice, usually with a diameter of less than 5 millimeters (0.2 in.).

Snow. Snow is precipitation that reaches the ground in the form of ice crystals. Sometimes single ice crystals fall to the ground, but more often several of them coalesce to form *snowflakes.* The size of snowflakes may vary from a few millimeters to several centimeters across.

Hail. Hailstones, which may range from 5 millimeters to 10 centimeters (0.2–4 in.) or more in diameter, are rounded lumps of ice that fall during thunderstorms. Composed of layers of transparent to translucent ice, hail forms where very strong updrafts are present in the thunderstorm clouds. Hail is a spring, summer, or autumn phenomenon that never occurs when ground temperatures are below freezing.

Types of Precipitation

Condensation and precipitation take place when moist air is cooled below its dew point. The onset of precipitation will depend largely on the air's moisture content and on how fast and how much it is cooled from its initial state. Large masses of air can be cooled quickly only if they are lifted to higher elevations rapidly, which lowers the pressure and allows the air to expand. Three methods of elevating air masses quickly include lifting of warm air over cold air (cyclonic); thermal lifting of an air column to form a thunderstorm cloud (convective); and lifting against a mountain front (orographic).

Cyclonic Precipitation. Cyclones are large air masses several hundred miles across with low pressure at the center and with winds blowing around the center in a clockwise direction in the Southern Hemisphere and counterclockwise in the Northern Hemisphere. Satellite pictures, shown in the weather section of many large daily newspapers, show cyclones as large cloud masses with spiral strands of clouds curving in toward the center. These systems, which are like large waves in the atmosphere, move in a general west-to-east direction. When they reach southerly moving polar air (in the Northern

Hemisphere), the boundary between the air masses forms a "front"; this is where most cyclonic precipitation takes place. Depending on how the air masses come together, the junctions are termed "warm" fronts or "cold" fronts. In either case precipitation occurs when warm air slides up and over a mass of heavier, colder air.

A warm front is formed when an advancing mass of warm air moves up an inclined surface of retreating cold air and is chilled in the process of being lifted to a higher elevation (see Figure 4.2). The lifting produces the cooling, not the cold air underneath. As a matter of fact, apparently little mixing goes on between air masses of different temperatures (and hence different densities), except possibly in the very turbulent zone near the ground. The zone of precipitation may stretch out for 200–300 miles (300–500 km) ahead of the front's surface location. Precipitation continues at a moderate to light intensity until the warm front passes on the surface.

A cold front forms when warm air is displaced and forced upward by an advancing mass of cold air (Figure 4.3). Here precipitation consists mostly of intermittent showers occurring near the surface front, but sometimes showers will fall 100 miles (160 km) or more ahead of the moving cold front. Cold fronts move faster than warm fronts and generally produce heavier precipitation, with the most intense precipitation falling near the surface front. After the front passes, clear, cold weather often results, particularly in the winter.

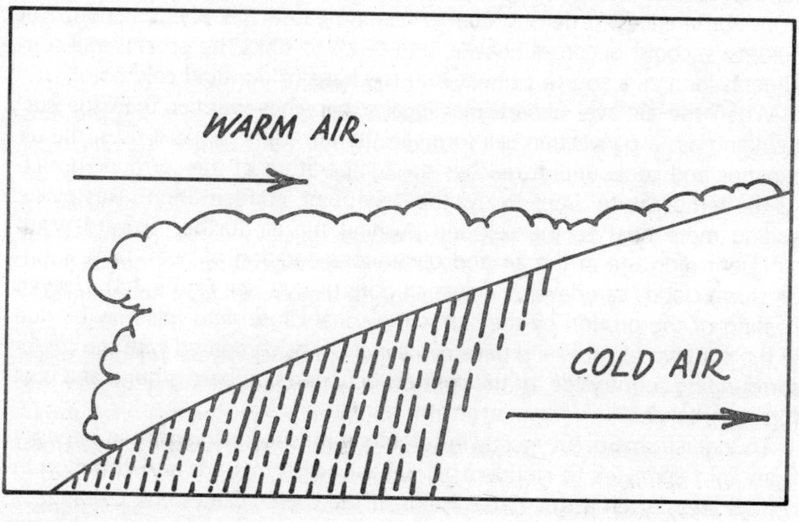

FIGURE 4.2 A warm front.

FIGURE 4.3 A cold front.

Convective Precipitation. *Convection* is the process of heat transfer from one place to another by the actual motion of hot liquids or gases. You may have noticed that the surface in a cup of hot coffee often is not still; it seems to move as though it were boiling; several small cells of liquid rise to the surface, overturn, and sink again. As the coffee cools, the motion (convection) stops. Differences in density due to heating cause this phenomenon. Hot fluid rises, cools, becomes heavier, and tends to sink. The process will continue as long as a source of heat is at the base of the fluid column.

When the air over a particular locality becomes warmer than the surrounding air, a convection cell forms in the atmosphere. As it rises, the air expands and cools until it reaches the temperature of the surrounding air. As the temperature drops below the dew point, condensation takes place, adding more heat to the cell and pushing the air further upward. With sufficient moisture in the air and continuous addition of heat, large thunderstorm clouds can develop in just an hour or two (see Figure 4.4). Uneven heating of the ground by the sun is the usual cause, and this may be due to the contrast of roofs and pavements in a city as compared with the cooler surrounding countryside, or the contrast between hot, bare ground and cool forests or lakes.

Thunderstorms occur mostly in summer with showers varying in intensity from light sprinkles to cloudbursts. Severe hailstorms are also common in certain areas, such as the Great Plains of the United States and Canada.

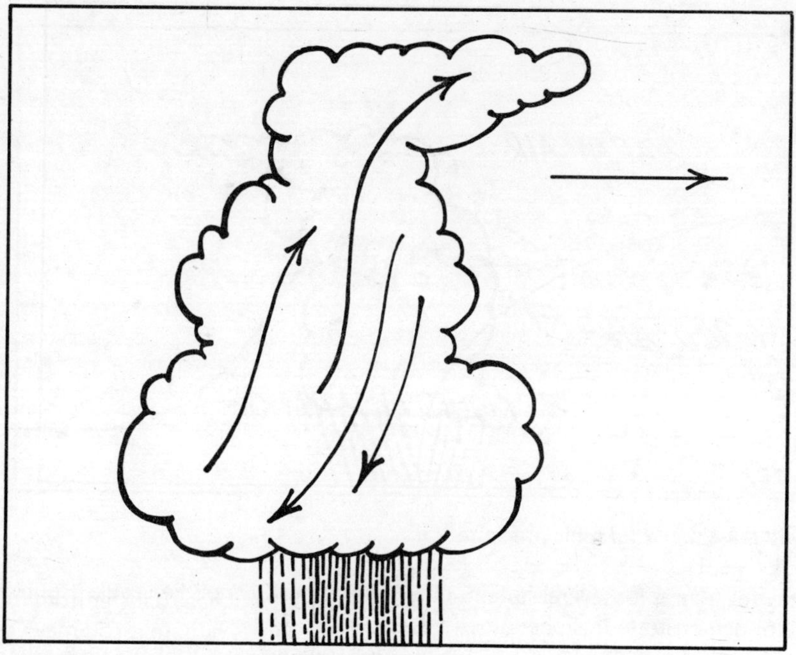

FIGURE 4.4 Convective thunderstorm cloud.

Orographic Precipitation. The term orographic is derived from *oros,* the Greek word for mountains. Orographic precipitation is a special type associated with the windward sides of the mountain range (see Figure 4.5), and it often occurs in combination with cyclonic or convective precipitation. This is usually a low-intensity type of precipitation, but the constant flow of maritime air against the ranges may raise the total annual rainfall.

Probably the best example of orographic precipitation in the United States occurs in the mountain ranges that extend along the Pacific coast states. A narrow band of high annual rainfall runs parallel to the ranges on their seaward side. Warm, moist air from the Pacific blows against this mountain barrier, lifting it high enough to cause cooling, condensation, and precipitation on the western flank of the mountains. As the air, now depleted of much of its moisture, passes over the mountains it flows to a lower elevation on the lee side, contracting in volume and consequently rising in temperature. Lower moisture and a higher temperature mean less condensation and precipitation. The lee side of the coastal mountains lies in a "rain shadow"; its annual precipitation is much lower than that on the windward slopes. Annual

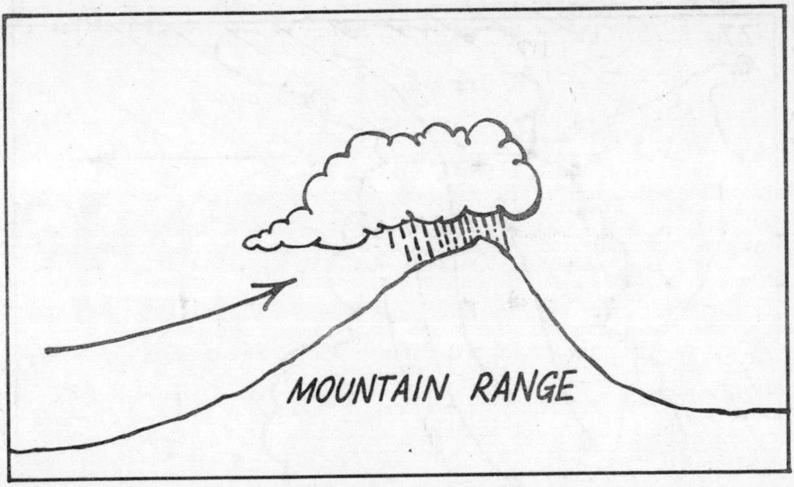

FIGURE 4.5 **Orographic precipitation.**

rainfall figures for several stations on Washington's Olympic Peninsula (Figure 4.6) demonstrate this phenomenon very well.

Even laws pertaining to water use are sometimes written to take into account the variation in rainfall on the two sides of the mountains. In Oregon for many years, laws governing groundwater use and conservation applied only to areas east of the Cascade Mountains. Lands on the west (seaward) side of the mountains presumably always had plenty of water from rainfall.

Because of its constancy and year-after-year reliability, orographic precipitation has influenced the establishment of human communities in many parts of the world. Most of the people in Oregon and Washington, for example, live on the west side of the Cascades. The balance of political power also has always been in the west. People in eastern Oregon and Washington have long felt neglected by their lawmakers in the state capitals, which, naturally, lie west of the mountains. Earlier in this century eastern parts of both Oregon and Washington began a strong movement to secede and form a new state. The state was to be called Lincoln, and even though many felt strongly about the idea, nothing ever came of the effort—because (you guessed it) not enough people would make up the new state! The uneven water balance will keep the preponderance of population on the rainy side of the mountains.

Measurement of Precipitation

Precipitation is measured in inches or millimeters of water. Any sort of open can or bucket with vertical sides can collect water, and sometimes after a

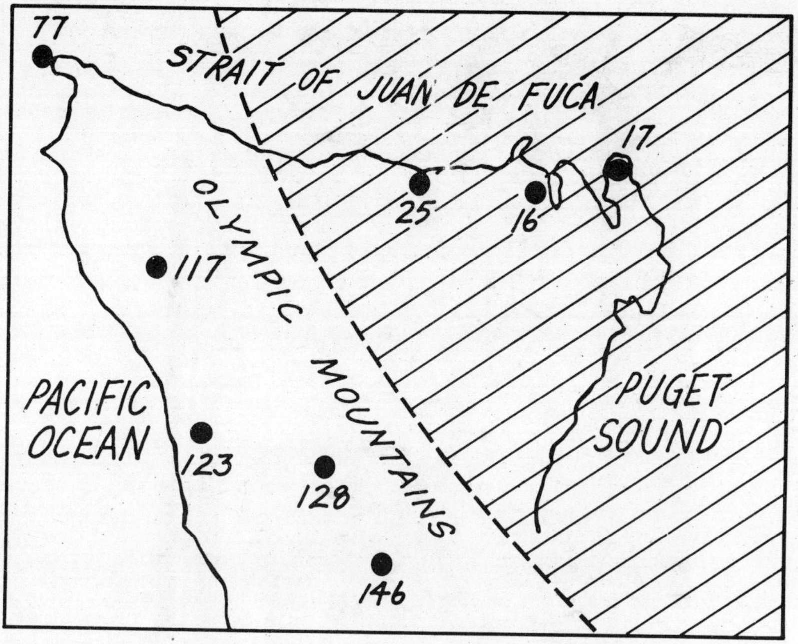

FIGURE 4.6 Variation in mean annual precipitation (inches) on Olympic Peninsula, Washington.

big storm hydrologists go through the countryside, measuring water depths in any sort of container they can find with water in it. For regular measurements at established stations, however, standardized gages are used. These differ from country to country, but their intent is to measure and record precipitation consistently and uniformly.

In the United States the National Weather Service uses a standard gage (shown in Figure 4.7) with an 8-inch (200 mm) opening at the top and a measuring tube that has a cross-sectional area 10% that of the upper 8-inch (200 mm) opening. The gage thus magnifies the depth for easy measurement; one inch (25 mm) of rain falling in the opening produces 10 inches (250 mm) of water in the measuring tube. Using a measuring stick graduated in inches and tenths, estimating rainfall to a hundredth of an inch is simple. Most European rain gages are similar, except that instead of measuring the water in the collector with a graduated stick, the collected rainwater is poured out into a graduated cylinder, of the type chemists use. Of course, the measurements are in millimeters rather than inches. When snow is expected, the observer can remove the inner measuring cylinder, let the snow collect in the overflow can, then melt the snow and pour it into the cylinder for measuring. These kinds of gages are usually read every 24 hours.

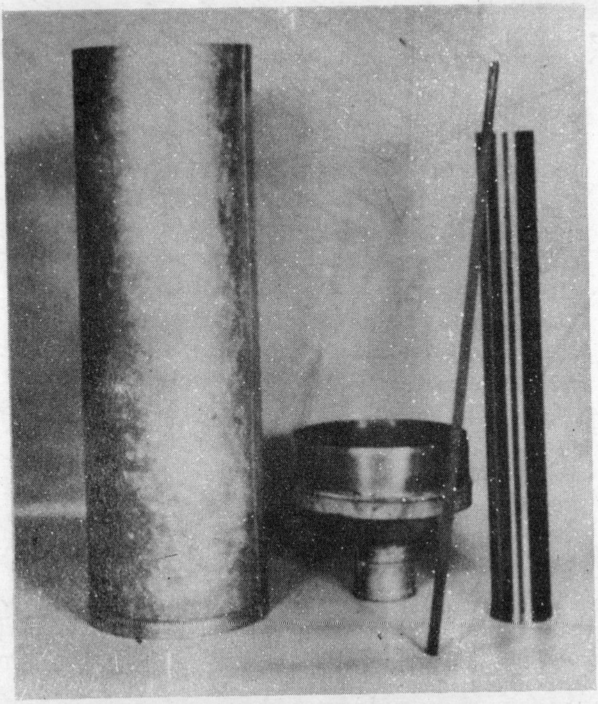

FIGURE 4.7 Standard nonrecording 8-inch (200 mm) precipitation gage (source: National Weather Service).

Another type of instrument that has come into widespread use in recent years is the recording gage. The chart from this gage will show not only the total rainfall in a given period but also the intensity—the rate of fall per minute, per hour, and so on. One common type has a collector bucket sitting on a scale that activates a pen on a moving recording chart (see Figures 4.8 and 4.9). These kinds of rainfall data are needed to develop design criteria for hydraulic structures such as culverts under roads, storm drains in cities, and so on. For example, data from a nonrecording gage might show one inch (25 mm) of rainfall collected in a 24-hour period. Data from a nearby recording gage might show that the entire amount fell in one 30-minute period during that same 24 hours. You needn't guess which gage record would be most useful to the engineer designing storm drains in that locality.

Recording gages usually would be preferred if not for their cost. They are more expensive than the simple nonrecording gages, however, and many more of the nonrecording type will always be in use. More than 10,000 nonrecording and 3,000 recording precipitation gages are in use in the United States.

FIGURE 4.8 **Recording, weighing-type precipitation gage (source: National Weather Service).**

Another means for studying rainfall makes no use of measuring gages. Radar, in the 1–20-centimeter wavelengths, shows rain as distinctive bright areas on the scanning monitor. This indicates to the meteorologist and hydrologist where the heaviest rainfall is in a storm and aids in interpreting data from the scattered gages within the storm area.

Location and Spacing of Gages. Precipitation is measured on a routine basis at all National Weather Service stations and at many other kinds of

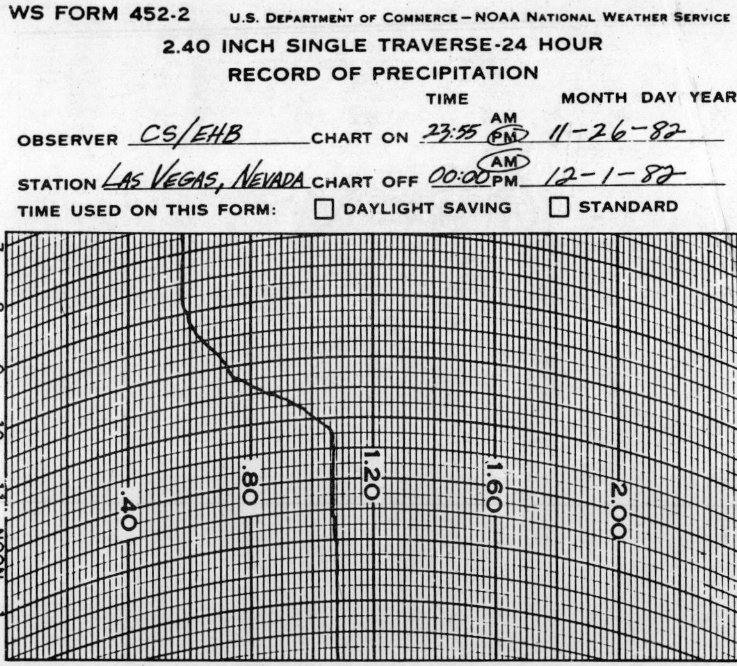

FIGURE 4.9 Chart from recording precipitation gage (source: National Weather Service).

U.S. governmental installations. All cities and nearly all the towns in the United States have an official weather observer or volunteer who measures rainfall and reports it to the National Weather Service. Between cities and towns, a multitude of people measure and record precipitation. Country people are always interested in the weather, and many families on remote ranches have for years maintained a continuous record of rainfall. A number of these private records have been collected and published by various state and federal agencies. In remote areas of the West some of these privately collected records contain the only precipitation data available. The map in Figure 4.10 shows the density of precipitation measuring stations in the United States.

The geographic location of a measuring station is usually dictated by circumstance. The gage must be read or the chart changed periodically, and that means that a human observer must visit the station. For remote locations where it is especially desirable to have measurements, however, some kind of telemetering scheme (wire or radio) can be used to transmit the readings back to a base station. These remote gages can be left unattended for up

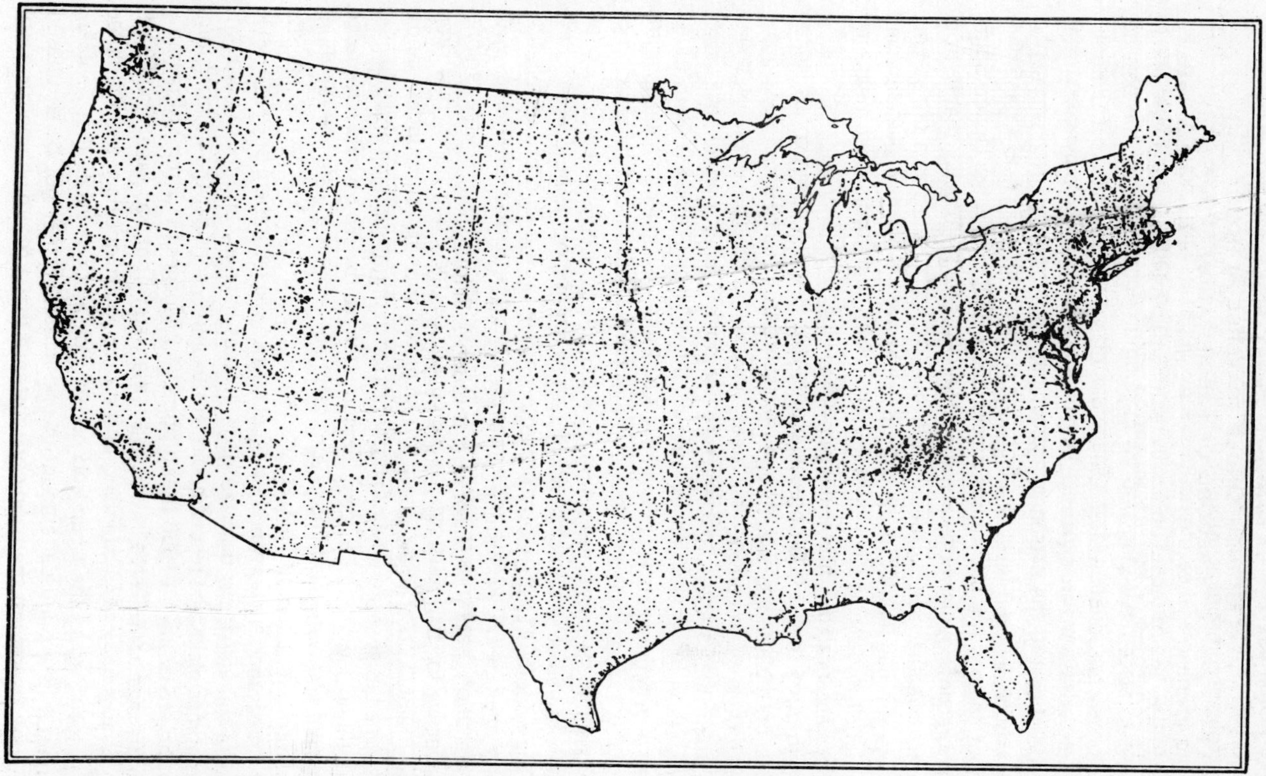

FIGURE 4.10 Locations where precipitation is regularly measured in the contiguous United States (source: National Weather Service).

to a year. In heavily populated drainage basins subject to severe flooding, telemetered rain gages often help engineers determine when heavy rainfall is apt to cause imminent flooding downstream. Remote gages of this type also provide periodic measurements of snow depth in high mountain regions.

As in the case of recording versus nonrecording gages, one might like to be able to place measuring stations at optimum locations for the best acquisition of precipitation data. The problem is once again cost. Remote, telemetered stations are expensive, and ultimately the hydrologist must justify the added cost to the person or agency paying the bills. Often a potential observer, living only a few miles from the chosen site, will read a gage daily without charge. The majority of precipitation stations in the United States are nonrecording, many operated by unpaid observers (unpaid, that is, from National Weather Service funds). They may be firefighters at a fire station, rangers in a national forest, or citizens who have a sense of civic duty and an interest in daily weather. In any case it is obvious that most measuring stations are located conveniently to people.

With more than 13,000 precipitation gages in the conterminous United States, on the average each station represents an area of about 235 square miles (600 km²). In considerable areas, however, especially in the West and in Alaska, that average spacing does not hold. Remote areas have fewer precipitation gages and densely populated areas have relatively many more. In the heavily urban areas of California, for example, about one gage operates for every 2 square miles (5 km²). In other parts of the world the spacing of precipitation stations varies widely; it is denser in Europe than in the United States, but some countries have only a few measuring points.

The importance of spacing depends largely on the kind of precipitation being measured. Cyclonic and orographic storms blanket large areas with precipitation that tends to be at least roughly uniform over sizable regions. Convective (or thunderstorm) precipitation, on the other hand, may drop heavy rainfall on a small area and completely miss the surrounding lands. Everyone has at some time or another stood in bright sunshine and watched dark thunderstorm clouds drenching an area nearby. You would have to be nimble if you wanted to move your rain gage around to catch all the thunderstorm precipitation. Instead, meteorologists establish a permanent site for a measuring station and hope that chance will favor them once in a while with a thunderstorm over the gage. It does happen; and those occasions sometimes make the record books. For example, consider the thunderstorm on the 4th of July in 1956 that dropped 1.23 inches (31.2 mm) of rain in 1 minute over a small area in Unionville, Maryland. That was a real cloudburst. If you had been there standing in the rain, you might have been tempted to "swim" up to the cloud and have a look around! While that rainfall is reported as an unusual amount in one minute, it may not be so unusual after all. It was just the happy circumstance that a recording rain gage was under that very pregnant thunderstorm cloud. Records like this one help us to understand signs along highways in the arid southwestern United States (where thunderstorms are common), warning travelers not to camp in "dry" stream beds. Southwestern natives call those kinds of storms real "gully washers," and with good reason.

Errors in Measurement. The purpose of a precipitation gage is to collect all the rain or snow that falls during a storm. Most instrumental errors cause a loss of some precipitation and result in a low record. When a gage is dry (between storms), an estimated 0.01 inch (0.002 mm) of rain is required to wet the collecting funnel and measuring cylinder surfaces before the cylinder begins to collect measurable amounts of water. Over a year, this could represent a loss of as much as one inch (25 mm) of measurable rainfall. Similarly, evaporation in a nonrecording gage that is read only once in 24 hours could cause a small loss of measurable water over a year's time. Dents in the metal rim around the top of the collector funnel may produce errors if they change the ratio of collector to cylinder area. Raindrop splashes from the collector funnel reduce the catch of the gage, particularly for large raindrops in a strong wind. Probably the most serious source of error is strong wind accompanying the precipitation. Wind passing over a gage exerts a certain amount of upward "lift," which can deflect precipitation and reduce the yield. The effect is increased for snow, prompting experiments with various kinds of shields to try to solve the problem. Shields tend to collect snow, however, providing another source of error.

The best way to minimize instrumental errors is through carefully locating the gage. In the past many weather service offices occupied tall buildings in the middle of cities, with measuring instruments on the roof. Wind conditions make these poor locations for rain gages and evaporation pans. Furthermore, temperature and humidity measured on a high rooftop are not typical of city environments. Airports now house many weather stations; there, the instruments can be located in more suitable surroundings. The best locations for precipitation gages are on level ground out in the open, far enough away from trees or buildings to prevent their interference with the gage catch, but close enough for them to serve as a windbreak.

Precipitation Patterns in Space and Time

Geographic Range of Precipitation. If you were asked "Why does it rain a lot in some places and hardly at all in others?" you might answer that it depends on their latitude and their relation to the circulation pattern of the atmosphere. In general you would be right. With abundant warm, moist air rising continually from tropical seas and flowing north and south, rainfall tends to be heaviest near the equator and to diminish as the air flows toward higher latitudes. What about two places at almost exactly the same latitude that are among the driest and wettest places on the earth?

Along 25° north latitude in the Western Desert of Egypt, the region around Kharga goes year after year with no rain at all. Its long-term average is probably 1 millimeter (0.04 in.) per year. At the same latitude in northeastern India a town called Cherrapunji holds the world record for the most rainfall in a 12-month period—1,041 inches (26,441 mm) in 1860–61. Its long-term average is 430 inches (10,922 mm) per year. While latitude is important, then, other factors must cause geographic variations in precipitation. For example, Cherrapunji is located on the edge of a plateau that provides the first orographic barrier to the moisture-laden winds of the monsoon blowing north out of the Bay of Bengal. Kharga, on the other hand, lies in a large depression near the eastern side of the great Sahara Desert, more than 2,000 miles (3200 km) from the Atlantic Ocean and about 500 miles (800 km) south of the Mediterranean Sea. By the time winds from the ocean reach Kharga they are as dry as the sands they blow across.

In addition to latitude, therefore, the geographical distribution of precipitation depends on orographic factors as well as how far an air mass has moved away from its source of moisture. Over North America, even though there is much evaporation from the land, the major source of precipitation is maritime air from the Pacific and Atlantic Oceans and from the Gulf of Mexico. The Great Lakes contribute some moisture, and cities on the downwind sides of the lakes have somewhat higher seasonal precipitation than those upwind, but these are minor factors in the total pattern of continental precipitation. The map in Figure 4.11 shows the complex distribution of precipitation over the United States. The decrease in rainfall going north from the Gulf of Mexico clearly shows the effect of air moving away from its source of moisture. The orographic effects of the mountain ranges in the West also appear clearly on the map.

Seasonal Variations in Precipitation. While no true seasons occur in the tropics (from about 10° N to 10° S), precipitation varies with the sun's position. Heavy rains are more frequent in September, October, March, and April, when the sun is almost directly overhead, supplying abundant heat energy to the atmosphere and ocean surface. Other parts of the world have more definite seasons around the year and widely varying precipitation patterns. Many parts of the world have one definite rainy season, which may occur in summer or in winter depending on the general circulation, prevalence of weather fronts and cyclonic storms, frequency of summer thunderstorms, and so on. Northern Australia has a rainy season in the summer and southern Australia has one in winter. Southern Africa has its rainy season mainly during the summer; and as we saw in the case of Kharga, Egypt, portions of northern Africa have no rainy season.

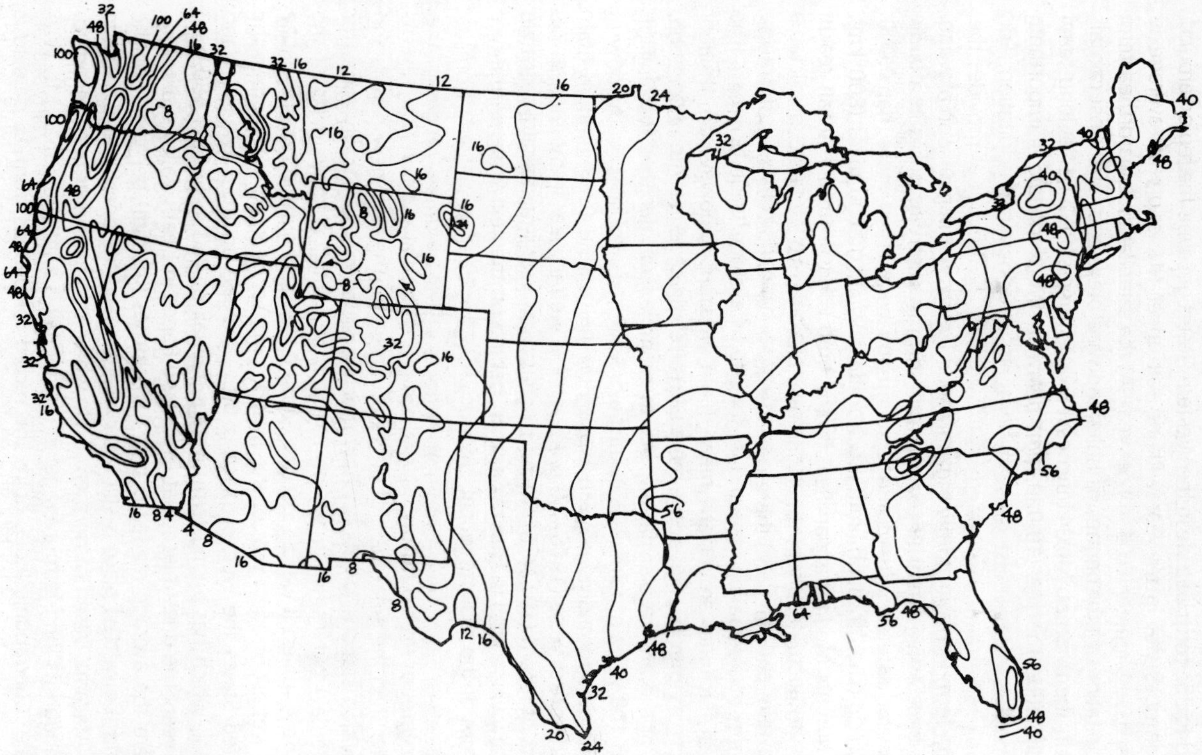

FIGURE 4.11 Mean annual precipitation in the United States (source: National Weather Service).

The United States, because of its varied terrain and position in relation to sources of maritime air, shows distinct regional patterns. As shown in Figure 4.12, Pacific coast and western mountain states get their main precipitation in winter. States in the Great Plains and upper Mississippi River drainage basin have a more even distribution, but with a tendency toward maximum precipitation in the summer. This is due to the prevalence of thunderstorms during the summer. Most states east of the Mississippi have fairly uniform precipitation the year around, particularly in the Northeast and northern Midwest.

Long-term Variations in Precipitation. The famous banker, J. P. Morgan, was once asked what he thought the stock market was going to do. He answered astutely, "It will fluctuate." If you were asked about the course of precipitation through time, you would be safe in giving the same answer, but you might want to qualify it. While both stock prices and precipitation seem to fluctuate randomly, stocks have exhibited no verifiable, predictable long-term pattern. Precipitation, on the other hand, does have a predictable pattern of fluctuation—around its long-term average value or mean. The way it fluctuates around the mean is not predictable, however. A series of wet years (precipitation above the mean) often precedes a series of dry years, but not in any regular pattern. As an example, consider the plot of more than 100 years of rainfall at Bakersfield, California, in Figure 4.13. Looking at the data for individual years plotted on the chart, you will note that one of the driest years (1976–77) preceded the wettest year in more than 100 years of record (1977–78). Trying to guess next year's rainfall at Bakersfield (or anywhere else) is even riskier than betting on the next throw of the dice or turn of the roulette wheel.

Extremes of Precipitation. News comprises unusual and significant events rather than the great mass of uneventful routine. The local newspapers and television stations will report how much it rained in your locality today (or yesterday) but usually make no mention of rainfall in other parts of the country—unless it was an extreme amount. Annual figures often emphasize the wettest or driest year or the wettest or driest geographic region. In the United States the wettest place is Mt. Waialeale on the Hawaiian island Kauai, which has a long-term average rainfall of 460 inches (11,680 mm) per year. The driest place is at Greenland Ranch in Death Valley, California, where the average rainfall is only 1.78 inches (45 mm) per year. For the world the wettest place is still Mt. Waialeale, and the driest is at Calama in the Atacama Desert of northern Chile where rain has never been recorded. Although its long-term average annual rainfall is a little less than Mt. Waialeale, Cherrapunji, India, holds the record for periods from 15 days to 2

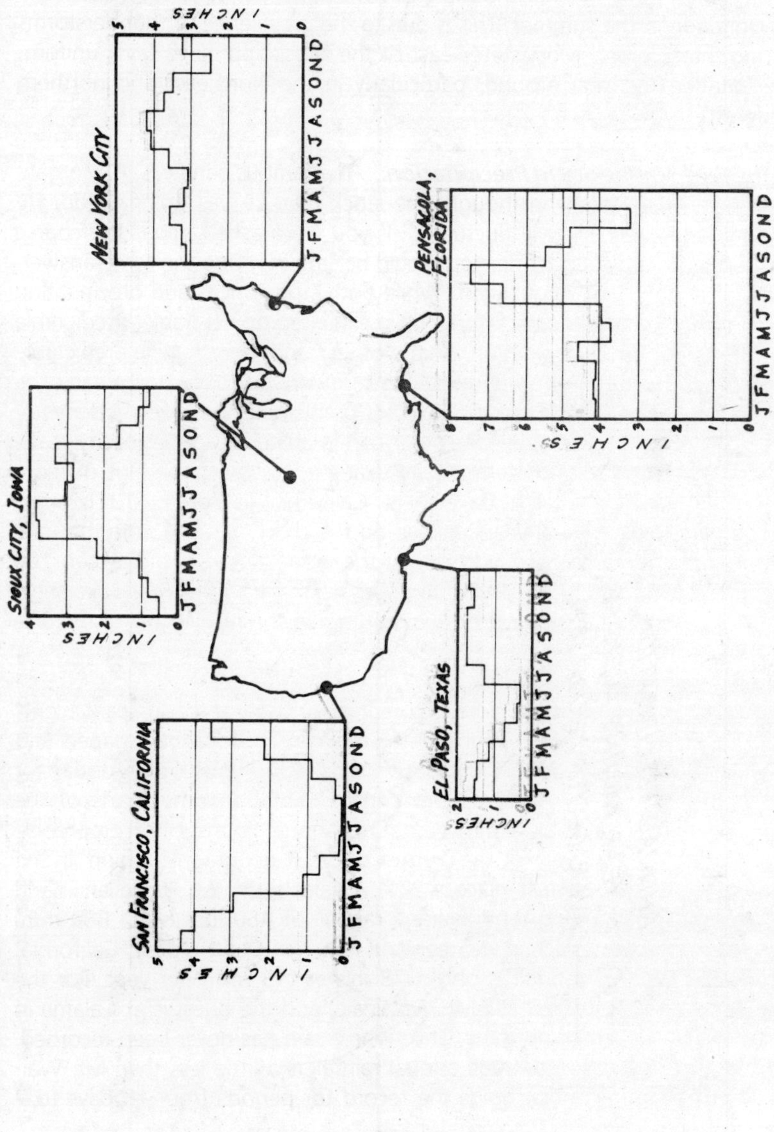

FIGURE 4.12 Regional precipitation patterns in the contiguous United States (data from the National Weather Service).

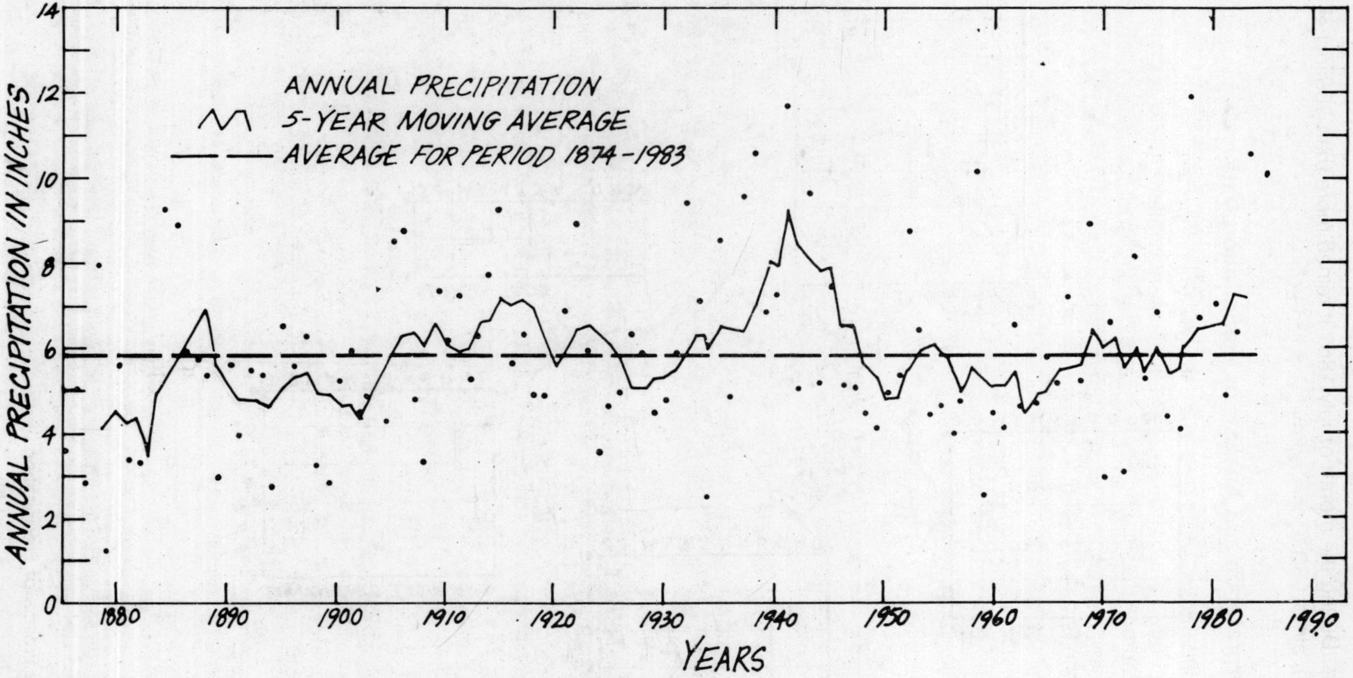

FIGURE 4.13 Precipitation at Bakersfield, California—1874–1983.

years. During the month of July 1861 it rained more than 366 inches (9,300 mm), almost 80% of a normal year's precipitation.

Interpretation and Analysis of Precipitation Data

Precipitation data find various uses. Weather reporters on the nightly television news announce today's rainfall or snowfall, then discard the figures; their interest is solely in current data. The ski resort operator with slopes still not covered by snow is very interested in daily snowfall figures until the ground has a sufficiently thick blanket to ensure good business. Farmers with growing crops watch weekly and monthly rainfall totals anxiously until enough rain has fallen to carry the crops through to harvest. Other users of precipitation records need information for various time frames. Hydrologists and engineers usually want the full range of data for the period of record, from the beginning of data collection through yesterday. Among other things, their concerns include long-term fluctuations in climate and weather patterns, especially the frequency and duration of unusually heavy precipitation.

The National Weather Service has done analytical studies on its very large store of U.S. precipitation records and has prepared many maps and charts to aid hydrologists and engineers in their work. The maps summarize precipitation events of various intensities and frequencies, indicating such things as average precipitation for spring, summer, fall, and winter; maximum annual precipitation; minimum annual precipitation; and more. A set of maps particularly useful for engineers designing drainage facilities summarizes intensity-duration-frequency data on precipitation for the entire country. These maps show how much rainfall one could expect during a 30-minute storm every 2 years, every 10 years, and every 100 years. Similar maps show how much rainfall to expect every 2, 10, or 100 years for 1 hour of rainfall, 6 hours, 24 hours, and so on. Figure 4.14 shows a 100-year 24-hour rainfall map. Comparing this map with the map in Figure 4.10, which shows average annual precipitation, indicates the value of these studies for drainage design. If the so-called "hundred-year storm" were to last for 24 hours, it would provide a sizeable proportion of total annual precipitation in just one day in many parts of the country.

The professional hydrologist, like all scientists, obviously needs a library of reference material in addition to the raw data bearing on the problem at hand. While these generalized analyses are useful, though, because each hydrological problem is unique, hydrologists must also use their own analytical procedures. How, for example, can the data measured at individual stations be extended over the larger, surrounding area?

Computing Areal Averages. While the ordinary citizen is satisfied to know simply how much it rained in the surrounding neighborhood yesterday, the

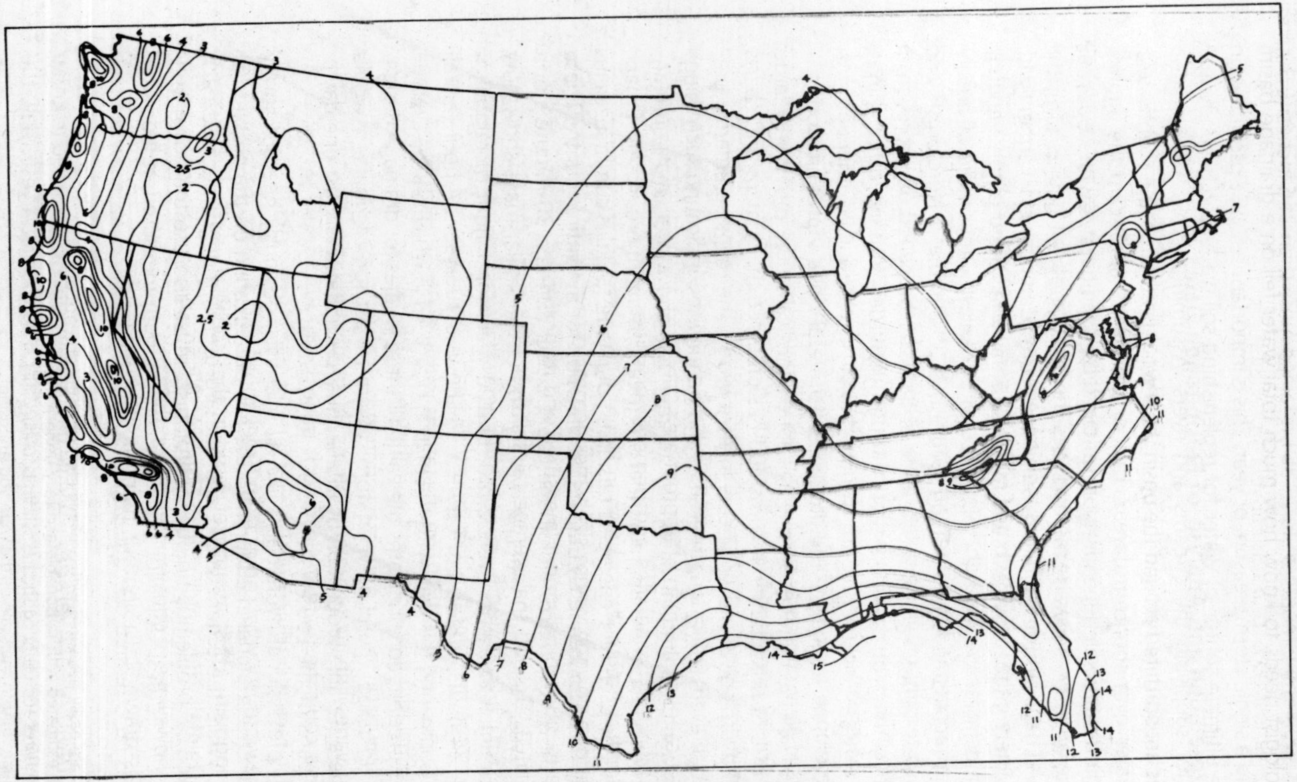

FIGURE 4.14 One hundred–year 24-hour rainfall for the contiguous United States (source: National Weather Service).

hydrologist needs to know how much total water fell on a drainage basin, during a single storm, season, or year. This is important because precipitation is the ultimate source of water for replenishing soil water and groundwater and for supplying the streams and rivers. Assuming several precipitation gages are scattered around the basin, how would we find the average depth of water over the entire area? This is possible using three more or less standard methods, depending on various local conditions within the basin.

The *arithmetic average* method uses the average of precipitation records from all gages in the basin. It is the simplest way of finding an areal average, and where gages are distributed fairly uniformly and produce approximately the same measurements, the method is as good as any other. While it would be best applied in flat country, the arithmetic method also works in mountainous terrain, provided gages have been properly located relative to orographic influence on storm patterns. Unfortunately, actual distribution of rain gages is generally not uniform across a basin.

Developed by A. H. Thiessen during the first decade of this century, the *Thiessen Network* has been widely used by engineers as well as hydrologists. It is a more formal method of computing areal averages than the simple arithmetic method, and it has the advantage of allowing for an uneven distribution of gages. A Thiessen network is made by first plotting all the gage locations on a map of appropriate scale. Next, straight lines are drawn connecting the gage sites and perpendicular lines are drawn through the midpoint of each line, as shown in Figure 4.15. Each gage is now near the center of a polygon whose size varies according to spacing between gages. The area of each polygon is measured and its percentage of the total area calculated.[1] To find the average rainfall over the basin, each gage precipitation total is multiplied by its polygon's percentage of area; all these figures added together produce the adjusted areal average rainfall. Besides correcting for uneven distribution of stations within the basin, this method also allows inclusion of data from some stations just outside the basin boundary. One drawback to using Thiessen networks is the assumption that precipitation varies linearly between stations. In mountainous country a Thiessen analysis could be quite misleading if gages were not located to take orographic factors into account.

The word *isohyetal* comes from Greek roots—*iso* meaning equal and *hyet* meaning rain. Thus, an isohyet, used in the *isohyetal method,* is a line of equal rainfall. The rainfall map in Figure 4.10 is an isohyetal map. An isohyetal map shows lines of equal rainfall drawn the same way a topographic contour map is drawn. In a contour map, various points of known elevation are

[1]Areas are easily measured with a planimeter, or in the absence of such an instrument, areas can be estimated with reasonable accuracy using a transparent overlay printed with a grid of small squares and counting squares within boundary lines. You would, of course, have to relate the square grid to the scale of the map you are using.

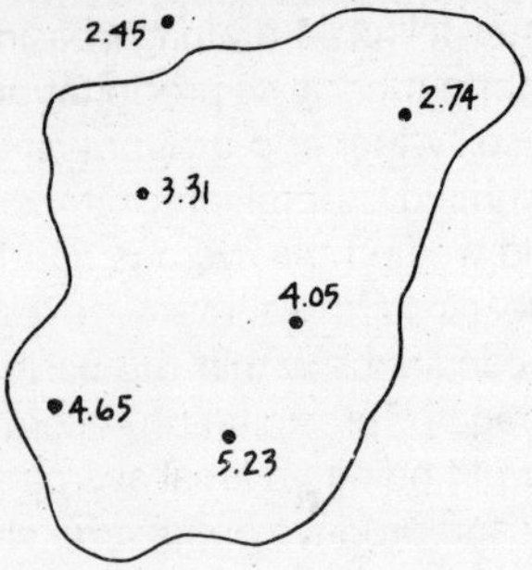

ARITHMETIC AVERAGE

$$\frac{2.74 + 3.31 + 4.05 + 4.65 + 5.23}{5}$$

AVERAGE 4.0 INCHES

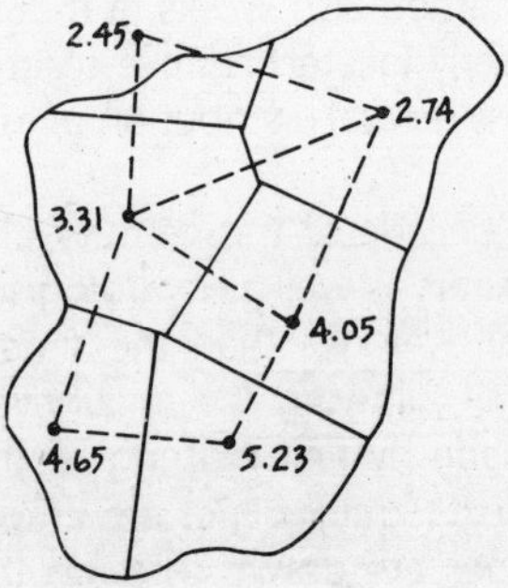

THIESSEN METHOD

1 STATION PRECIP. (INCHES)	2 AREA IN BASIN (SQ. MILES)	3 PERCENT OF TOTAL AREA	4 ADJUSTED PRECIP. (Col. 1 X Col. 3)
2.45	25	4	.10
2.74	123	22	.60
3.31	115	20	.66
4.05	113	20	.81
4.65	85	15	.70
5.23	106	19	1.00
	567	100	3.87

AVERAGE = 3.87 INCHES

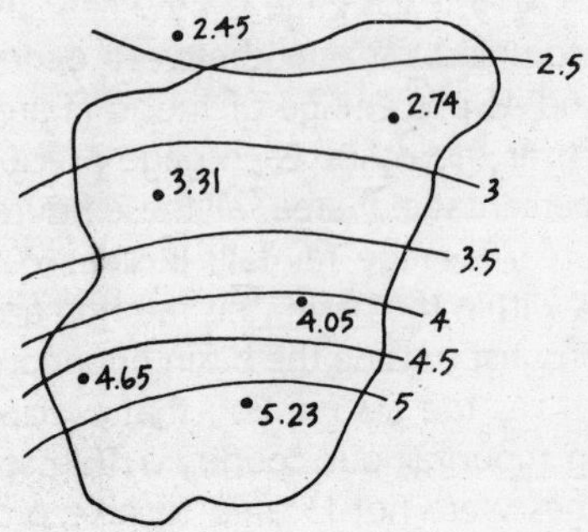

ISOHYETAL METHOD

1 AVERAGE ISOHYET (INCHES)	2 AREA BETWEEN ISOHYETS (SQ. MILES)	3 AREA X AVG. ISOHYET (Col. 1 X Col. 2)
≤ 2.5	29	73
2.75	117	322
3.25	120	390
3.75	73	274
4.25	51	217
4.75	68	323
≥ 5	109	545
	567	2144

$$\text{AVERAGE} = \frac{2144}{567} = 3.78 \text{ INCHES}$$

FIGURE 4.15 **Methods for computing an areal average for precipitation.**

plotted and lines of equal elevation (contours) are drawn using these known points as controls. Every map has a contour interval—a difference in elevation between contours. Similarly, an isohyetal map would have a rainfall interval between isohyets—one inch, two inches, and so on (or 10 millimeters, 20 millimeters, 30 millimeters, etc.). The sketch in Figure 4.15 shows how an isohyetal map might have been constructed using the same data from the arithmetic average and the Thiessen network. Average precipitation over the whole area could be determined as follows:

1. Find average precipitation between isohyets. This would be the arithmetic average of the two isohyets; between the 1- and 2-inch isohyets the average would be 1.5 inches, between 2-inch and 3-inch isohyets the average would 2.5 inches, and so forth.
2. Find the area on the map between two isohyets.
3. Multiply the area between isohyets by the average precipitation between them.
4. Find the sum of figures obtained in step 3 and divide it by the total area of the basin to obtain the average depth of precipitation over the basin.

The isohyetal method is the most accurate of the three, but it depends heavily on the skill of the person drawing the isohyets. The mapmaker should know the terrain within the basin and be able to interpolate the isohyets to truly reflect natural conditions in the field. In mountainous country the isohyets would approximate topographic contours, since orographic precipitation usually increases going upslope toward the mountain summit. Also, if some measuring points show unusually heavy precipitation, the isohyets can be clustered closer around such points. In the hypothetical examples shown in Figure 4.15 the averages for the three methods just happened to be close together. Ordinarily we might expect more variation in averages if we were using actual data from a real drainage basin.

There are a number of other ways to analyze precipitation data, but they are generally beyond the scope of this book. For further information on the subject of analysis, the interested reader is referred to the professional hydrology texts listed in Appendix IV.

SNOW

The foregoing sections of this chapter have used the term *precipitation* as including all forms of water falling from the sky to the earth; in reality, however, only two forms are widespread enough to be routinely measured: rain and snow. About 30 inches (760 mm) of precipitation falls on the average each year over the entire United States. Only about 4 inches (100 mm) of this average amount falls as snow; the rest is rain. Of course, not every reporting station receives both rain and snow. Some places have only rain and some have mainly snow (Figure 4.16). The southern states and the low

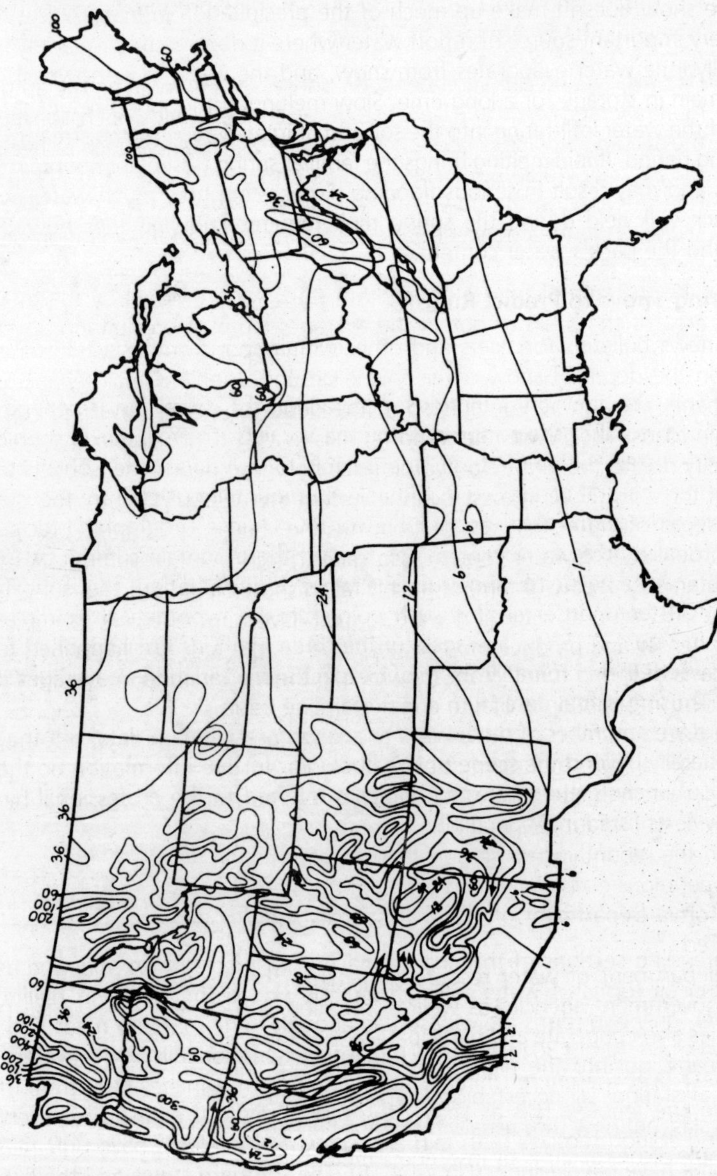

FIGURE 4.16 Mean annual snowfall (inches) in the United States (source: National Weather Service).

deserts get very little snow, while the high mountains get most of their precipitation as snow.

While snow doesn't make up much of the precipitation in many places, it is a very important source of runoff water where it does occur in quantity. Generally little water evaporates from snow, and the water in a snowpack may remain in storage for a long time. Slow melting in the spring results in much of the water infiltrating into the soil and ultimately feeding the streams for a long period. Rapid melting brings the familiar spring freshet in mountain streams and may result in serious flooding. Forecasting how much water a snowpack will yield during the spring melt requires knowing the snow's depth and the pack's water content.

Measuring Snow to Predict Runoff

Special news bulletins for skiers and other winter sports enthusiasts often report on the depth of snow cover on the ground. Highway maintenance departments are also very interested in the depth of new snow covering mountain roads. The TV weather reporter tells us that 6 inches of snow fell on the city during the night. To hydrologists, however, depth of snowfall is only half the story. Their interest is in the *water equivalent* of the snowpack, or the depth of water that results from melting the snow. Because snow varies in density, the water content also varies. Fresh snow averages about 10% water, so 2 feet (610 mm) of freshly fallen snow would produce about 2.5 inches (63 mm) of water if it were to melt. Over time a snowpack may become denser and produce more water.

Forecasts of spring runoff from snowmelt in the mountains have a variety of uses. The information is vital to engineers responsible for forecasting and tracking floods on rivers at the base of the mountains. Knowing how much flow to expect from mountain streams is important to power companies with hydroelectric generating stations along the streams. Runoff forecasts also provide important information for irrigation districts with upstream reservoirs in the mountains for seasonal storage of irrigation water. The snow survey operations each winter and spring in the Sierra Nevada mountains by the California Department of Water Resources show how this information is gathered.

The department of water resources manages the program, but several federal government agencies as well as irrigation districts and public utility companies also cooperate and participate in the actual surveys. In midwinter, survey teams go into the high Sierra to measure snow depth and water content at stations along established traverses or courses. Historically they went on skis, but now they usually fly in on a helicopter. At measuring points they sample a core of snow by inserting a metal tube vertically into the snow, then weighing the tube and snow on a special scale calibrated to indicate the snow's water content (Figure 4.17).

FIGURE 4.17 Snow survey party making an observation at a measuring point along an established snow course in the Sierra Nevada. Man on right is inserting tube into snow to obtain core. Man in the middle is holding the special spring balance used to weigh tube full of snow. The balance is calibrated to read in inches-of-water-equivalent in the snow (photo by California Department of Water Resources).

In addition to the snow courses traversed on the ground, a number of depth gages are checked periodically from the air, (see Figure 4.18). An increasing number of remote, telemetered gages send back information on snow depth by wire or radio to central data-collecting stations.

FIGURE 4.18 Snow depth gage read from the air (photo by California Department of Water Resources).

Ordinarily the department conducts four surveys during the winter and spring, publishing resuts in special bulletins on the first of February, March, April, and May. Normally the snow begins to melt about the first of April, and by that time state hydrologists are able to issue runoff forecasts for operators of water-storage reservoirs, hydroelectric power stations, flood-control engineers, and irrigation districts. The forecasters try to predict not only total runoff from the mountains but also the rate of runoff. This can have great value to those concerned with downstream use and control of water.

Armed with the runoff forecasts, a company like Pacific Gas and Electric Company (PG&E) can plan more efficient operation of its electric power system. PG&E has a large hydropower capacity on Sierran streams, and if it can expect a good runoff season, it won't need to buy as much oil or gas to run steam power generators during the summer. All major streams draining the western side of the Sierra Nevada (where most of the runoff occurs) have been dammed to store water for use and to prevent floods downstream. The engineers in charge of flood control can use the runoff forecasts to operate storage reservoirs, leaving space to catch excessive runoff and prevent floods. Irrigation districts serving farms with runoff water during the spring and summer can let the farmers know how much water to expect during the season and hence how many acres of crops to plant.

Often an interesting conflict arises between the flood controllers and the irrigators. The irrigators want storage space filled to the brim with water that they can use as needed. The flood controllers worry about an unusual warm spell, or warm rain on the snowpack, that will melt the snow faster than expected; they insist on keeping a safety margin at the top of the reservoirs. Accurate forecasts based on snow surveys allow conflicts like these to be resolved rationally.

Finally, consider how many snowflakes it takes to produce all that runoff. According to a forecast bulletin of the California Department of Water Resources,

> Snowflakes—The powder snow type—weigh only about 0.06 milligrams, which means that it would take 7.5 million of the crystals to make a pound. There can be as many as 10 million snowflakes in a cubic foot of snow, but to get a cubic foot of water from these snowflakes would require melting 468 million of them and to get one acre-foot of water would require—20,386,080,000,000 snowflakes. To fill Lake Oroville would require—[even more].

INTERCEPTION

If you have ever run for cover under a large tree when it began to rain, you have experienced interception in action. The leaves or needles of the tree "intercepted" the rain drops and kept them from falling on you or on the ground. Interception catches more of the precipitation in forested areas than in grasslands, and most water is caught at the beginning of a storm.

At first glance, overall water loss to interception would appear to be negligible, but this is often not the case. Most storms are of short duration, and interception takes the first drops to fall. If only a small amount of rain falls, vegetation may intercept a fairly large proportion and it will never reach the ground. In a forest with a dense canopy of trees, as much as 10%–20% of the annual precipitation may be lost to interception. Interception "steals" a certain amount of water whenever vegetation of any kind covers the ground.

The first raindrops in a storm remain on vegetative surfaces as a thin film of moisture, held against gravity by hydrogen bonding with the leaf surface. Only after all leaves in the path of the raindrops are wetted will rain begin to roll off the leaves and fall to the ground. During a prolonged storm, water eventually flows steadily down the stems or trunks of the plants to the ground.

Even during a rainstorm, the air near the ground is rarely saturated and considerable water can evaporate from the wet leaves. This might seem to be a trivial amount, but if you consider how much surface area the thousands of leaves or needles on a large tree represent, it is apparent that the total evaporative surface of the leaves is very large. A 10%–20% loss of annual precipitation doesn't seem out of line at all. In many light, short rainstorms, as much as 80% of the storm water could be intercepted and evaporated. Anyone who has taken cover from rain under a single tree or a forest canopy in such a shower stays fairly dry under the tree. A more prolonged rain will ultimately penetrate the tree cover, getting those beneath it wet.

Although the process is obvious and clearly results in some loss of precipitation, it is seldom possible to make more than a general estimate of water lost. Interception measurements have been done on an experimental basis for several types of vegetative cover, but useful data for solving most hydrologic problems are scarce.

Even though interception may deprive us of some precipitation, it is valuable because it greatly reduces the impact of raindrops on the soil. This effect helps maintain an open pore structure at the soil surface, thus promoting infiltration and inhibiting runoff and erosion. The results of removing the vegetative cover are obvious almost every year in the Los Angeles area. In their undisturbed state, the coastal mountains of Southern California are covered by a very dense growth of shrubs called *chaparral.* These are extremely flammable, and almost every year a summer brushfire burns off large areas of steep hillsides near Los Angeles. When winter rains begin, with little vegetation to reduce the amount and impact of falling water, the pores in the soil surface soon clog, infiltration decreases, runoff increases, and large mud slides often result. Interception, as a natural process, has a valuable role to play in the water cycle; what we lose in total precipitation probably is more than compensated for by the benefits of soil (and ultimately, water) conservation. As discussed in the next chapter, the interception of rainfall by grasses or crop debris may also play a crucial role in preserving the fertility of U.S. farmlands.

APPLICATIONS

Weather Modification

Mark Twain purportedly remarked: "Everyone talks about the weather, but no one does anything about it." That's not to say people haven't tried to do something about it. For centuries, primitive societies living in arid lands have conducted religious ceremonies trying to bring rain during a drought. The Indian rain dance is well known. In more recent times many Europeans believed that shooting cannons or rockets into the clouds could produce rain or suppress hail. During the nineteenth century itinerant "rainmakers" firmly established themselves in U.S. folklore. All these attempts at weather modification, however, produced many more failures than successes, since no one really understood what happened in a cloud to cause precipitation.

In 1946, scientists at General Electric's laboratory at Schenectady, New York, discovered something that resulted in an important breakthrough for the science of meteorology. They found that dry ice (solid CO_2), dropped into a cloud containing supercooled water droplets, can cause precipitation to fall from the cloud. As the small pellets of CO_2 fall through the cloud, they lower the air temperature enough to form ice crystals directly from water vapor. The supercooled cloud droplets immediately begin to evaporate and recondense on the ice crystals, which eventually become large enough to fall from the cloud as snowflakes. When air near the ground is warm, the snowflakes melt and fall to earth as rain.

Encouraged by their success with CO_2, the scientists experimented with other substances and eventually found a chemical—silver iodide—that would also cause precipitation from clouds. Silver iodide, which has a crystalline structure almost identical to that of ice, works differently than the CO_2, but is equally effective in producing precipitation nuclei in supercooled clouds. Instead of being simply dropped into a cloud, as CO_2 pellets are, silver iodide is dissolved in a flammable liquid, then vaporized in a gas flame to form a "smoke" of billions of tiny silver iodide crystals. When introduced into a supercooled cloud, the silver iodide smoke "seeds" the cloud and causes ice crystals to form, with the resulting precipitation similar to that caused by CO_2. Silver iodide has one marked advantage over CO_2: The CO_2 must be dropped into the clouds from an airplane, while the silver iodide can be introduced into clouds either from an airplane or from a ground-based smoke generator. Today most weather-modification projects employ silver iodide for seeding clouds. Figure 4.19 shows a typical remotely controlled silver iodide generator. The close-up view in Figure 4.20 shows the generator in full operation.

The research at General Electric in the mid-1940s did more than show how to increase precipitation from clouds. The overall results of the research greatly increased our knowledge of cloud physics and the general nature of the precipitation process. The same process the scientists were able to induce with dry ice or silver iodide occurs naturally and is one of the chief mechanisms for producing precipitation. The General Electric scientists simply gave nature a gentle nudge and caused precipitation from a cloud that was almost ready to precipitate anyway. This is a very important point to remember; clouds won't produce precipitation unless conditions are just right.

After results of the General Electric experiments were published, many would-be rainmakers took to the field with silver-iodide generators in an effort to bring rain. Following several years and many failed attempts, enough

FIGURE 4.19 Silver iodide smoke generator operated jointly by the U.S. Bureau of Reclamation and the Desert Research Institute of the University of Nevada. The tanks on the ground contain the liquid fuel (propane) and the silver iodide solution. Located at an elevation of 5,300 feet (1615 m) in the Sierra Nevada range west of Lake Tahoe, the generator was used in a cloud-seeding experiment aimed at increasing precipitation and runoff in the headwaters of the Truckee and Carson Rivers in western Nevada. The generator was remotely controlled by radio from an operations center at the University of Nevada in Reno (photo by U.S. Bureau of Reclamation, 1972).

FIGURE 4.20 Close-up view of generator shown in Figure 4.19. The gas flame, seen in the inner chamber, is vaporizing a silver iodide solution and sending a stream of silver iodide "smoke" up into the atmosphere (photo by U.S. Bureau of Reclamation, 1972).

knowledge accumulated to indicate when cloud-seeding operations could succeed. Now the cloud seeders don't fire up their silver iodide generators until they know the clouds are ready. One of the main things they determine first is the target-cloud-mass temperature.

Silver iodide crystals are effective as precipitation nuclei at cloud temperatures between −5° C (23° F) and −30° C (−22° F). Clouds with temperatures much below −30° C probably already contain an abundance of ice nuclei; adding more through cloud seeding is likely to reduce rather than increase the chances for precipitation.

Other criteria besides temperature determine successful seeding operations—amount of liquid water in the cloud, number of natural freezing or condensation nuclei in the cloud, atmospheric stability, and so on. Because it is often a rough index of the other favorable or unfavorable conditions and because it is relatively easy to measure, however, cloud temperature is the main criterion for beginning the seeding operation. To measure the temperature, a field meteorologist merely sends up a radiosonde and records the results. A radiosonde is an instrument carried aloft by a helium-filled balloon that senses temperature, humidity, and pressure, and broadcasts the data back to the earth.

This brief discussion is only a bare outline of cloud-modification methods. The reader wanting to dig deeper into this subject should refer to Appendix IV.

Cloud Seeding in Practice

With the abundance of newspaper accounts of cloud seeding over the years, most people just assume it is a routine operation performed wherever additional precipitation is desired. In reality, even today many weather modification projects are still at least partially experimental. The reason is that the amount of increased precipitation due to cloud seeding has been hard to determine with certainty.

The problem arises because most seeding operations are not started until precipitation is about to begin, or has already begun, in the target cloud mass. Depending on thermal conditions within the clouds, the seeding may either increase or decrease total precipitation, as measured by rain gages on the ground. The seeding could conceivably have no real effect at all on total precipitation.

After seeding a pregnant cloud and measuring the resultant precipitation, then, how can we determine how much of that precipitation was due to our intervention? Without it, would the measured precipitation have been much different? Meteorologists analyzing results from some of the early seeding experiments recognized this dilemma and tried to overcome it by planning experiments so they could analyze the results using statistical techniques. Essentially this required randomly choosing clouds to be seeded. If nature operated its "rain machine" on a random basis, the scientists reasoned that they would have to do the same.

Guidelines for designing experiments that produce randomized data are similar in all fields of natural science. If possible, a control group or area is chosen for comparison with the treated group or area. Cloud-seeding experiments thus require selecting a control area with meteorological characteristics similar to the treated area. Both areas are equipped with similar networks of rain gages, and the control area is so situated that the seeding will not affect it.

To guard against any possible bias (conscious or unconscious) by the person who controls seeding operations, generally a set of random decisions is made before field operations begin. Flipping a coin, using a set of random numbers, or another chance method produces the decisions which are placed in sealed envelopes to be opened in sequence as seeding operations progress. Instructions in the sealed envelopes would include the kinds of actions contemplated for the ensuing project: They might read "Seed," "Do not seed,"

"Fire up the northerly set of generators," "Fire up the westerly generators," and so forth.

Although plenty of uncertainty still surrounds cloud seeding, meteorologists using a random-decision method can apply standard statistical techniques in analyzing the resultant precipitation data. Statistical methods will never reveal the exact precipitation increase or decrease, but when applied to random data, they will produce at least a ball-park estimate of the seeding's effect. Although some researchers have made extravagant claims for large precipitation increases, the consensus among long-time workers in the field indicates that seeding produces average increases in precipitation of about 5%–10%.

Cloud seeding research has been carried on in the United States for many years, both by government agencies and private companies. Some of the most fruitful research has been that by the U.S. Bureau of Reclamation (USBR) and the Pacific Gas and Electric Company of California.

Project Skywater, managed by USBR, is a national research effort that employs university scientists and private consultants in addition to a permanent staff of government scientists. Because of the need for water in the West, most research has been carried on in the western states. The major emphasis has been on increasing precipitation from orographic storms in the western mountains and from summertime convective clouds over the Great Plains.

One of the most ambitious Skywater projects was the Colorado River Basin Pilot Project, operated for five seasons in the early 1970s in the San Juan Mountains of southern Colorado. Analysis of the results, which required almost two years to complete, showed that cloud seeding both increased and decreased snowfall. It was concluded, however, that results of the research and good future management of seeding operations might be used to increase seasonal snowfall by as much as 10%. If true, seeding winter storms in the mountains around the Colorado River Basin could add as much as 1.3 million acre-feet[2] (1,603,550,000 m^3) in additional runoff to the Colorado during an average year. This much additional water in the river could produce 1.07 billion kilowatt hours of hydroelectric energy from the generators at Glen Canyon Dam and Hoover Dam, estimated as equal to about 1.7 million barrels (270,278 m^3) of oil. In addition to the extra energy, the extra water would be a welcome dividend for many water users along the lower river. If applied in an operational project, the USBR has estimated that direct benefits from such a cloud-seeding program would probably exceed costs by at least a 10 to 1 ratio.

One of the longest-running cloud-seeding programs in the United States (and probably in the world) is one by the Pacific Gas and Electric Company

[2]Acre-foot: The volume of water required to cover 1 acre to a depth of 1 foot; equivalent to 325,851 gal. or 1233.5 m^3.

(PG&E). For almost 30 years PG&E has conducted cloud-seeding activities on the western slopes of the Sierra Nevada range. The company has tried to increase runoff in the rivers draining the western slope, thereby increasing the power produced by hydroelectric plants situated on the rivers.

PG&E is a pioneer utility company in California that serves an area stretching from the southern San Joaquin Valley to the state's northern borders. The company and its predecessors built many of the existing hydroelectric plants along the Sierran streams. Hydropower originally provided a substantial proportion of the company's electrical energy. As demand has increased with the population growth in recent years, the company has built steam-electric plants to supply additional power. For the most part these plants burn oil or gas, which the company must purchase. If PG&E can enhance runoff on mountain streams, then, it will save on alternative fuel sources. In addition, the people downstream can use the additional water for recreation, irrigation, and domestic purposes.

Cloud Seeding at Lake Almanor.[3] One of the oldest and best-documented cloud-seeding programs in the world is the one run by PG&E at Lake Almanor in California (see map in Appendix I for location). Situated along the boundary between the Cascade range to the north and the Sierra Nevada range to the south, the Lake Almanor watershed covers about 500 square miles (1295 km^2) of mountainous terrain just south of Mt. Lassen. Lake Almanor's elevation is 4500 feet (1372 m) above sea level, and although the watershed extends up the southerly slopes of Mt. Lassen, most of it lies below 7500 feet (2286 m). The wet season runs from November to May, with a substantial number of orographic storms during this period. Annual precipitation averages more than 30 inches (762 mm), with snowfall averaging about 146 inches (3710 mm) per year.

The first cloud-seeding experiments were performed in the early 1950s; as with most other experiments at that time, the results were inconclusive. In 1960 PG&E decided to change its experimental design and conduct the seeding randomly. During 1960 through 1962, eight radio-controlled, silver iodide generators were installed on ridge tops and more than 50 heated precipitation gages were placed in a 300-square-mile (777 km^2) target area of the watershed. Two control areas upwind of the target area were also equipped with heated precipitation gages. The experiment continued for five seasons (November 1 to May 15) from 1962 to 1967. Analysis of the data collected over the five seasons showed a positive effect from seeding for certain storm types: orographic storms with westerly winds and cloud temperatures of −5° C (23° F) *below* 7500 feet (2286 m) that PG&E calls *cold westerly storms.* Since 1967, PG&E has randomly

[3]Material for this case history of seeding operations at Lake Almanor was supplied by Margaret Mooney, former Director of Meteorology Services, and C. A. Threlkeld, Supervising Hydraulic Engineer, Pacific Gas and Electric Company, San Francisco.

seeded cold westerly storms with the requisite cloud temperatures. The company estimates this has yielded about 5% more precipitation over the watershed than would have occurred in the absence of seeding.

Another major storm type affecting the Lake Almanor area occurs with southerly winds and cloud temperatures at −5° C (23° F) *above* 7500 feet (2286 m); PG&E refers to it as a *warm southerly storm.* After seeding of the cold westerly storms was demonstrated to be effective, the experiment was redesigned to study warm southerly storms. By the end of the 1975 season, sufficient precipitation data had accumulated for a statistical evaluation of the warm southerly storms. This evaluation indicated that seeding in these storm types may have decreased precipitation. This apparent decrease could have been caused by overseeding, which may have produced so many condensation nuclei that few had a chance to grow large enough to cause precipitation. The decrease could also have been caused by other undetermined physical processes. Due to the significance of this finding, warm southerly storms are no longer seeded as part of the project. Two other classes of storms, *warm westerly* and *cold southerly,* continue to be evaluated as part of the Lake Almanor project.

The goal of PG&E's cloud seeding program is to increase the mountain snowpack, thereby increasing runoff and generating more electricity in the six power plants on the Feather River downstream from Lake Almanor. The additional runoff from seeding the cold westerly storms is estimated to increase annual power generation by about 1%. This amounts to about 30 million kilowatt hours of electricity, the equivalent of about 50,000 barrels (8000 m³) of oil or 300 million cubic feet (9 million m³) of natural gas that generating the power in a steam plant would have required. Since the cost of the cloud seeding program is just a small fraction of the cost of that much oil or gas, PG&E probably will continue it indefinitely.

As we learn more about it, we may be able to increase efficiency of seeding and enhance the benefit-to-cost ratio. A natural question might be: "If you can make some money by increasing runoff a little, couldn't you make a lot of money by increasing runoff even more?" Unfortunately the answer is no. Because hydroelectric power plants are so costly to build, their sizes depend on average river flows. A little more flow will produce a little more power, but when flow exceeds the plant's capacity, the water must pass the powerhouse and the energy is wasted. Seeding is thus done only in years when snowpacks are at average levels or below. In 1983, for example, when the snowpack was far above average, seeding operations were shut down.

Environmental Effects of Weather Modification

In the early days of weather-modification experiments people were concerned about the effects on their supply of natural precipitation. Were these cloud seeders wringing all the water out of clouds before they reached downwind areas? Would they dry up the natural storm systems and leave drought conditions in their wake? This concern was understandable. Rain shadows in the lee of western mountain ranges were well known. Would the modern-day rainmakers create new rain shadows in the lee of their seeded areas? The extravagant claims of greatly enhanced precipitation due to seeding only heightened the concern. Early proponents, for example, talked of drawing down large amounts of water from the "rivers in the sky." Furthermore, politicians' tendency to regulate anything they don't understand caused state legislatures to begin passing laws regulating weather modification.

Ultimately, though, even the most ardent cloud seeders weren't able to affect natural weather systems much. When the weather scientists studied precipitation patterns in and beyond the seeded areas, they had enough trouble trying to prove increased precipitation from seeding in the target areas, let alone changes in natural precipitation in the regions beyond. This is not surprising, considering the enormous volume of moisture in the air moving across the continent every day. Natural precipitation mechanisms are not very efficient in removing moisture from the air in any one storm, and a new supply of moisture-laden air is almost always moving in from the coast. After all the years of seeding experiments and analyzing results, the weather community now seems to agree generally that while a few arguments about overall seeding effectiveness remain, cloud seeding does not produce water in one region at the expense of another.

Another environmental concern is the residual silver iodide that comes down with rain or snow and then remains on the ground or flows away in the streams. If you were to drink water that had been precipitated with silver iodide, should you worry about silver poisoning? The answer is no, not in the least. The amount of silver iodide added to the atmosphere during seeding is very small, and not every silver iodide crystal forms a raindrop or snowflake. By the time the precipitation reaches the earth, the concentration of silver iodide in the water is extremely small. The USBR estimates that snow melting from Project Skywater seeding contains about one ounce (28 gm) of silver iodide in 500 million gallons (1,893,000,000 L) of water, roughly 0.000015 milligrams per liter. This is far below the 0.05 milligrams per liter limit for silver in drinking water set by the U.S. Public Health Service.

SUMMARY

Cloud seeding seems to be one place we can tamper with a natural system without damaging the environment. As the PG&E program in the Sierra Nevada showed, seeding seems to produce truly beneficial results in some places. The method is particularly attractive in regions with inadequate water supplies. In Israel, for example, cloud seeding is one of several methods used to enhance the country's meager water supply.

The countries of southern Africa have recently experienced the worst drought in half a century; water supplies have been depleted everywhere. The Republic of South Africa, with its rapidly growing population and increasing demands for water, probably has the worst shortage problem. The republic's Water Research Commission has estimated that demand for water will outstrip the supply by the year 2020. In addition, the quality of water is deteriorating from increased use and reuse of available supplies. Rainfall stimulation is one of the few alternatives for providing more water without adding dissolved salts that must be removed. The South Africa commission contracted with a firm of U.S. weather consultants to establish and manage an extensive, long-range research program aimed at increasing precipitation and eventually increasing the overall water supply. If the research succeeds, the results would benefit all of southern Africa.

5

Infiltration and Soil Water

. . . as water spilt on the ground . . . which cannot be gathered up again. . . .

(2 Samuel 14:14)

INFILTRATION

Infiltration is the flow of water from the ground surface down into the soil. An important part of the water cycle, it supplies most of the water to plants and many animals. Without continuous infiltration, wells would go dry, streams would stop flowing soon after a rain, farm soil would erode, and frequent floods would devastate the river valleys. Infiltration is a vital natural process.

What Is Soil?

Because the description of infiltration refers to water moving "down into the soil," a definition of *soil* might help here. In general, soil is an unconsolidated aggregate of mineral and rock fragments ranging in size from tiny clay and silt particles to sand and sometimes even to pebbles and boulders. To the soil scientist or farmer, soil is confined to the upper few feet near the surface, the zone occupied by the roots of growing vegetation. For example, soil scientists in the U.S. Department of Agriculture typically map soils to a maximum depth of 5 feet (1.5 m) on the assumption that practically all plant roots (except trees) will occur within this depth below the surface.

Where soil is relatively thin over the parent bedrock from which weathering formed it, soil scientists and hydrologists would describe the entire soil zone in a similar way. Where the same kind of porous, granular material extends for many feet below the root zone, though, the soil scientists and hydrologists would differ in their view of what constitutes soil. The hydrologist would follow the civil-engineering definition that considers soil to be all of the unconsolidated, granular material—including organic material—overlying hard bedrock. The thickness of "hydrological" or "engineering" soil could thus range from one foot to thousands of feet. One test well, drilled in Owens Valley, California, on the eastern side of the great Sierra Nevada fault scarp, bottomed in unconsolidated gravel ("soil?") at a depth of 8,000 feet (2438 m). This admittedly unusual depth is due to rapid sinking of Owens Valley along the fault, but unconsolidated sediments several hundred feet thick over bedrock are not uncommon.

Zones of Subsurface Water

As defined here, soil could contain both kinds of subsurface water, *soil water* and *groundwater* (see Figure 5.1). The well in Owens Valley ran out of the soil-water zone and into the groundwater zone within a few dozen feet of the surface. The rest of the hole was in the groundwater zone. How, then, do these two zones differ? Are they defined in some academic, arbitrary way? Not at all. They are fundamentally different because of the way water occurs in the two zones. The groundwater zone is also called the *saturated* zone, because water completely fills all the soil or rock pores. The top of this zone is called the *water table.* Soil water, on the other hand, occurs above the water table in the *unsaturated* zone where some pores contain water and some are essentially empty, containing only air. As shown in Figure 5.1, the terms used for describing zones of subsurface water are those of the U.S. Geological Survey. The unsaturated zone is the environment discussed in this chapter; Chapter 7 discusses the saturated zone.

The unsaturated zone functions in two ways: It acts as a storage reservoir for soil water, and it provides pathways for water moving down to recharge the groundwater or moving up toward the ground surface and the atmosphere. On the surface, or down in the saturated zone, water moves mainly through gravity, and the motion is always along a path of decreasing gravitational potential energy.[1] That's a technical way of saying that water flows downhill. This is no surprise because our experience with water in everyday life bears it out. It may come

[1]Gravitational potential energy is a relative term that denotes increased or decreased separation between an object and the center of mass of the earth (as stated formally in Newton's Universal Law of Gravitation). When an object is raised (moved away from the earth's center), it gains in potential energy by the amount of work required to lift it to its new position. As a matter of convenience, gravitational potential energy is often referenced to sea level, which is arbitrarily assigned a zero value. This is convenient because land surface elevations are also referenced to sea level, and it is easy to assign relative values of potential energy to objects based on their elevations above sea level. Thus water from rain falling on land at 5,000 feet (1524 m) above sea level has more potential energy than water from rain falling on land at 2,000 feet (610 m).

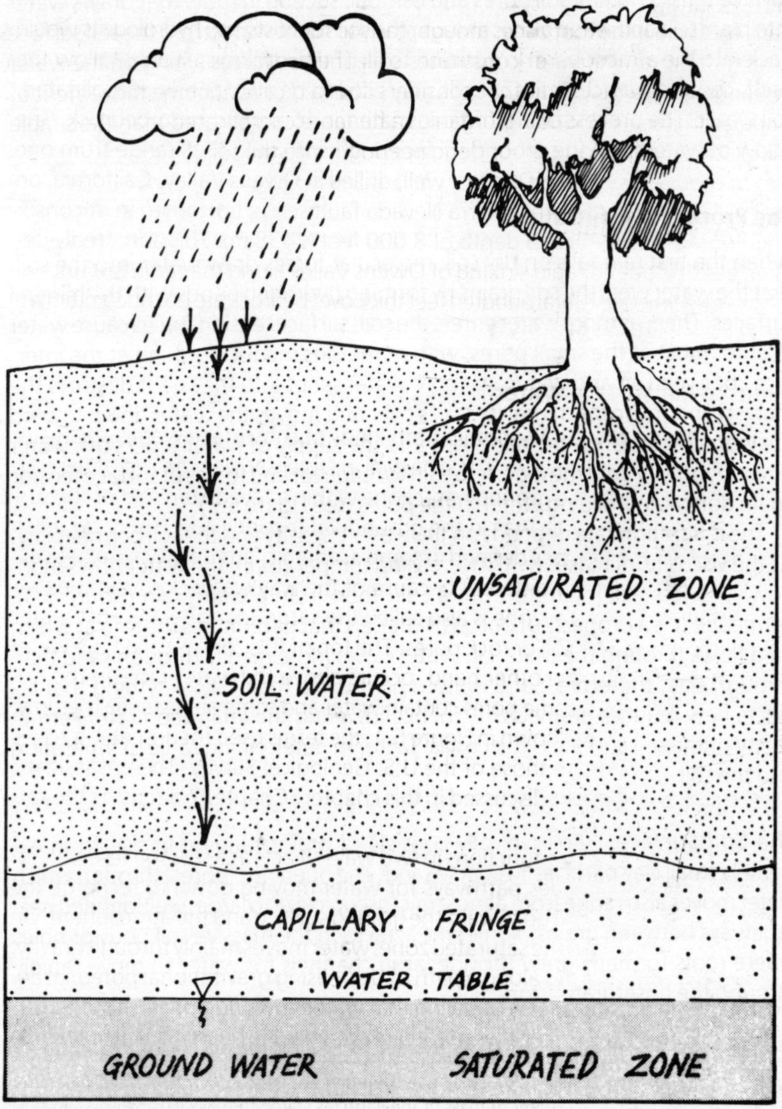

FIGURE 5.1 Zones of subsurface water.

as a surprise, however, to learn that in the unsaturated zone (above the water table) water may move in any direction—up, down, or sideways. In addition to gravity, water in this zone is subject to forces of capillary suction produced by surface tension in tiny soil pores and osmotic suction in roots that draws water into plants. Evaporation also is always ready to move water from the soil surface back into the atmosphere. In addition to all of these forces acting on the water itself, certain characteristics of soils may either aid or restrict the movement of soil water. The process of infiltration is a bit more complicated than the simple "flow of water from the ground surface down into the soil."

The Process of Infiltration

When the first rain falls on dry soil, molecular forces draw water into the soil. First the water wets the soil grains by forming hydrogen bonds with the mineral surfaces. Then, as more water enters the soil, surface tension forces cause water to enter most of the small pores, with numerous menisci forming at the interfaces between air and water. As more water seeps into the soil, it destroys the menisci and the force of gravity takes over and moves the water down toward the water table. Once the rain stops and the supply of water tapers off, water drains by gravity until the menisci begin to reappear and forces of surface tension again dominate, holding water in the pores (see Figure 5.2).

The infiltration rate depends on the water viscosity and the soil permeability. As the viscosity increases (with decreasing temperature), the water flows more slowly, so less will infiltrate from a cold than from a warm rain (Figure 5.3). *Permeability* is a property of porous materials that determines the rate at which fluid flows through the material. Water flows more readily through sand than through clay; therefore, sand is more *permeable* than clay. Anything that restricts flow reduces permeability, hence lowering the infiltration rate. Permeability of the soil surface is the crucial factor in determining rates of infiltration into the soil.

The Soil Surface. Soil typically consists of grains ranging in size from extremely small clay particles to coarse sand. The openings (pores) through which water moves also range from almost microscopic size to large, well-defined passageways between grains. Sometimes cracks in the massive soil body or holes where roots formerly grew provide large channels for water to move rapidly through the unsaturated zone.

The soil surface is a complex system of organic as well as inorganic components that undergoes almost constant change, both physically and chemically. Enlargement or reduction in the sizes of the interconnected pores that transmit water through the soil affect infiltration. Other changes also enhance or reduce permeability and infiltration. In certain situations, a simple change in water quality can reduce good infiltration rates to almost zero in a very short time. Intense rainfall also commonly reduces permeability in surface soil and hence reduces infiltration rates. The impact of raindrops compacts the soil surface.

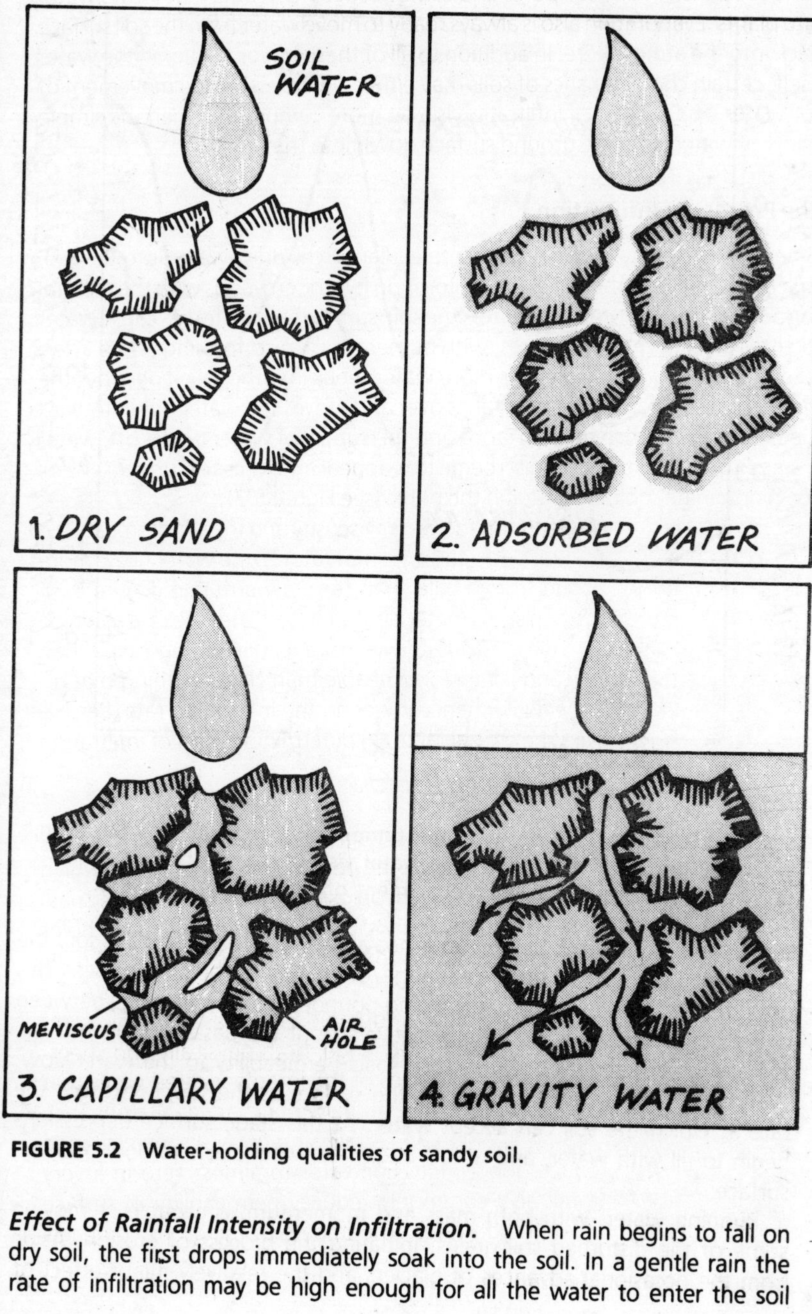

FIGURE 5.2 Water-holding qualities of sandy soil.

Effect of Rainfall Intensity on Infiltration. When rain begins to fall on dry soil, the first drops immediately soak into the soil. In a gentle rain the rate of infiltration may be high enough for all the water to enter the soil

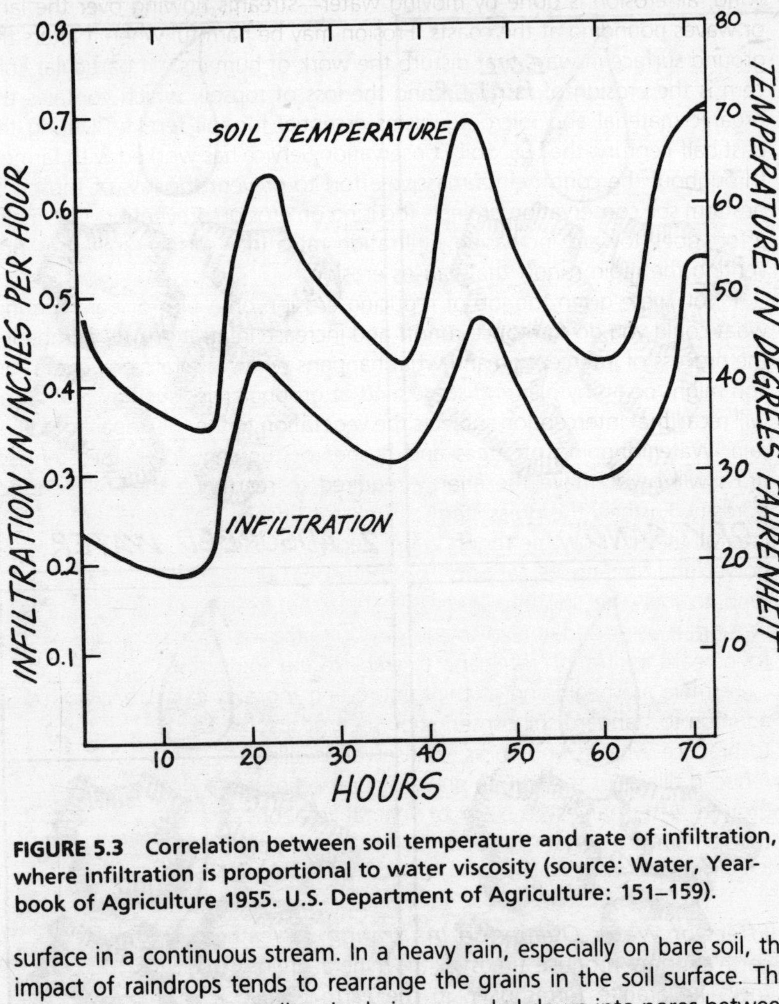

FIGURE 5.3 **Correlation between soil temperature and rate of infiltration, where infiltration is proportional to water viscosity (source: Water, Yearbook of Agriculture 1955. U.S. Department of Agriculture: 151–159).**

surface in a continuous stream. In a heavy rain, especially on bare soil, the impact of raindrops tends to rearrange the grains in the soil surface. This often results in many small grains being pounded down into pores between larger grains, thus reducing the effective size of the passageways into the soil and greatly reducing the surface soil's permeability to the water flow. Water may then begin to collect on the surface as the supply exceeds the rate at which the soil can accept water. At this stage surface depressions begin to fill with water, and runoff begins as water starts to flow over the surface.

Running water, with both mass and momentum, is capable of moving some of the disturbed soil grains, initiating the process of erosion. Aside from the occasional advances of glaciers and the relatively minor effect of wind, all erosion is done by moving water—streams flowing over the land or waves pounding at the coasts. Erosion may be harmful when it alters the ground surface in ways that disturb the work of humans. Of particular concern is the erosion of farmland and the loss of topsoil, which contains the organic material and microorganisms essential for soil fertility. During the past half-century, the U.S. Soil Conservation Service has worked with farmers throughout the country in a massive effort to prevent the loss of topsoil by on-farm soil conservation projects focusing on erosion prevention. The major effort goes toward increasing infiltration rates into surface soil, thus preventing the rapid runoff that causes erosion.

If you were given the job of reducing erosion on a plot of bare ground, what could you do to reduce runoff and increase infiltration? Remembering the process of interception and what happens when rain falls on vegetation, you might begin by planting some kind of ground cover such as grass. You will recall that interception subjects the vegetation to the full impact of falling rain. Water dripping off trees and bushes or running down the stems of grass will hardly have the energy required to rearrange the soil's surface porosity. Further, the grass stems, trunks of bushes and trees, and ground litter all tend to slow the runoff water, reducing its momentum and carrying power and limiting its capacity to erode soil. Roots of the vegetation also tend to lock the soil into a massive unit that resists disaggregation and erosion. If you decided to plant grass, you made a wise choice in beginning to increase infiltration rates and to stabilize the soil surface in your plot.

Farmers are beginning to think more and more in this direction, too. In addition to standard soil conservation methods, some farmers are beginning to practice what is known as "conservation tillage." This method calls for reduced tillage of the soil (to prevent compaction) during planting and crop growth. After harvest, a cover of vegetative debris is left on (or just under) the soil surface. Again, the purpose is to promote infiltration and reduce opportunities for erosion.

Effect of Water Quality on Infiltration. Weathering processes at the earth's surface produce almost all the mineral grains in soil. When a crystalline rock like granite forms deep in the earth's crust, it is stable in a high-temperature, high-pressure environment. When massive earth forces lift the granite and expose it at the surface, weathering begins. Most minerals that were stable in a deep environment are no longer stable under atmospheric conditions. The major exception is quartz—silicon dioxide (SiO_2)—which makes up most of the sharp, gritty grains in soil. Most other minerals begin to change as soon as they are subjected to water and gases in the atmosphere.

One of the most interesting results of the weathering process is the group of minerals called clays; these, along with quartz, are the main constituents of surface soil. The more than one dozen kinds of clays each have distinctive compositions, but all are composed of extremely small, individual mineral grains. Many are not much bigger than large molecules, and because they have a greater surface area for a given mass, clays are more chemically active than most large-grained minerals. Clay particles generally carry a negative charge and thus tend to attract positively charged ions. Positive ions are called *cations* because in a solution they are attracted to the negative pole, or cathode; negative ions are called *anions* because they are attracted to the positive electrode, or anode. As you will remember from Chapter 2, many natural compounds ionize, or separate into positive and negative ions when they go into solution in water. Certain of the common cations, namely calcium (Ca^{++}) and sodium (Na^{+}), play an important part in causing structural changes in soil aggregates that affect water infiltration and movement in soils. Equally important is the tendency of some clay minerals to swell when immersed in water.

When clay minerals undergo hydration, or the adsorption[2] of water into the crystalline structure, the mineral increases in volume (see Figure 5.4). Sometimes the volume increase is extreme. Montmorillonite, a common clay in many soils, may swell up to 15 times its original dehydrated volume. It is easy to see that when clays are near the soil surface and rainwater causes them to begin hydrating, tiny pores may fill with swelling clay. In cases like this the water blocks its own path into the soil, and infiltration decreases as runoff increases.

Sometimes the way water affects clay depends on the type of ions held in solution in the water and adsorbed on the surfaces of the clay minerals. Even though the individual clay particles are extremely small, they often combine to form aggregates called *crumbs*. These are held together by electrical forces in the clay and in the adsorbed cations. For example, clay crumbs bound with Ca^{++} ions form an aggregate with the texture of fine sand, despite the fact that individual clay crystals are so small they are invisible except under an electron microscope. When clays combine to form crumbs, they are said to be *flocculated*. They then form permeable aggregates that allow water to move through the soil. Sometimes raindrop impact will break up these crumbs, but with a vegetative cover protecting it, such a clay configuration may be stable for long periods. The structure of clay crumbs can be chemically disrupted, however, drastically reducing infiltration.

Introducing water containing Na^{+} ions may cause *deflocculation* of the clay and destruction of the crumb structure. This occurs when Na^{+} ions are exchanged for Ca^{++} in the clay mineral, destroying the electrical forces binding the clay particles together. Deflocculation of clays reduces infiltration

---

[2]*Adsorption* is the adhesion of the molecules of a gas, liquid, or dissolved ions to a solid surface. Water that wets a surface and is held against gravity is *adsorbed* on the surface. *Absorption* is the assimilation of molecules or ions *into* a solid or liquid substance, thereby forming a solution or compound.

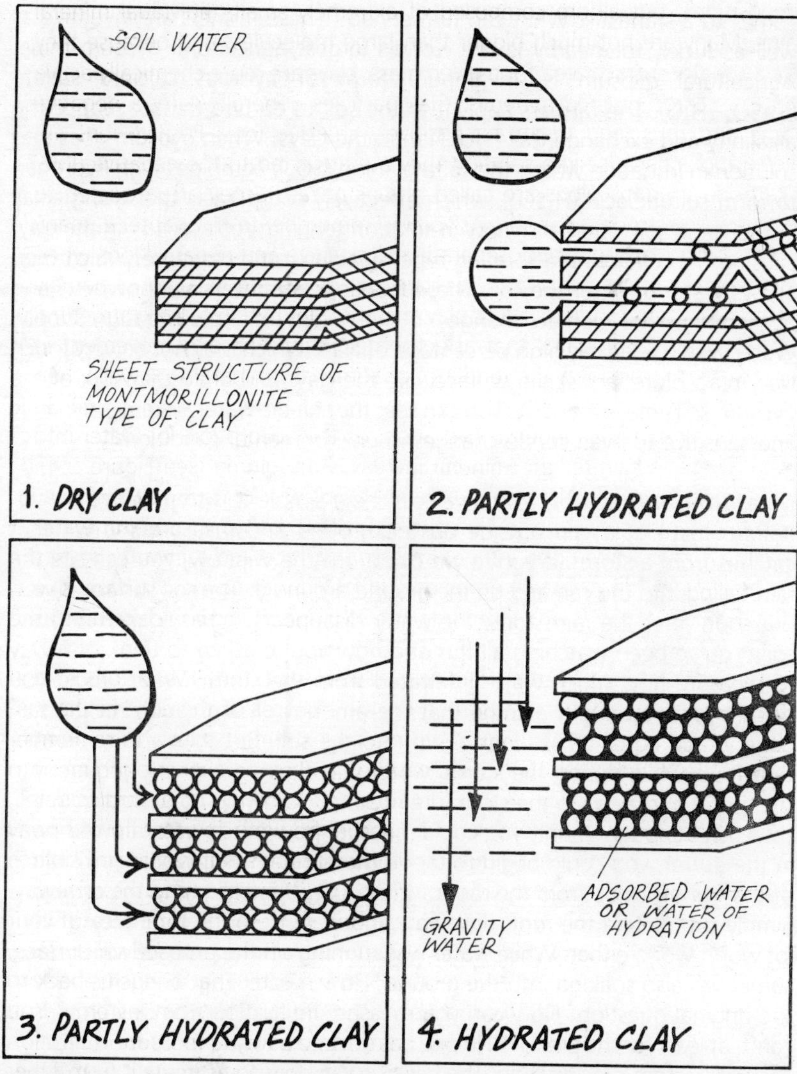

FIGURE 5.4 Hydration (and swelling) of clay mineral in soil.

and can frustrate farmers' efforts to get irrigation water to penetrate soil in their fields.

On the west side of California's San Joaquin Valley (see Map in Appendix I), many of the soils have large amounts of alkali containing abundant Na^+. Irrigation water applied to the surface often tends to run off, and farming comes to a standstill. What can farmers do? These farmers are inventive as well as lucky, for nearby is the world's largest agricultural gypsum mine. Agricultural gypsum is an impure form of hydrous calcium sulfate ($CaSO_4 \cdot {}_nH_2O$). The farmers apply it to the soil to do two things: reduce the alkalinity and exchange Ca^{++} for Na^+ in the clays. When gypsum goes into solution in irrigation water, the sulfate anion ($SO_4^=$) combines with hydrogen to form sulfuric acid (H_2SO_4), which effectively counteracts the strong alkali in the virgin soil. The Ca^{++} now in solution tends to replace Na^+ in the clay, causing flocculation and forming a permeable crumb structure, which then allows water to enter the soil. These reactions are reversible, however, and farmers must be alert to changes in the quality of their irrigation supply. Water with a relatively high percentage of Na^+ (which may not be very much Na^+ in absolute terms) can deflocculate the clays and undo the work of the gypsum soil amendment. As you can see, the soil-air-water system is dynamic and sensitive to even subtle changes in environmental conditions.

Measuring Infiltration. You've seen how simple it is to measure precipitation. Just put a can outside on the ground and measure the water it catches from a storm. If you're watching out the window, you can see the rain falling into the can and on the ground around it. The soil surface is wet, but soon after the rain stops, the water disappears. It has soaked into the soil. You've been watching all this and now you're asked to think of a way to measure how much water infiltrated from the storm. What would you do? At first you might assume that the amount of rain caught in the rain gage would equal what went down into the soil. But then you remember that for a while during the storm, water was flowing over the ground into the gutter and away to the storm drain. Simply knowing how much rain fell, then, isn't enough. If only you could measure how much water flowed away in the gutter, you could begin to tackle the problem. But would you? Much of that runoff came from the roof of the house, the walks, and the driveway. Simply subtracting the roof, driveway, and walk from the total area of your lot won't work, either. While water was running off the exposed soil surface, some was also soaking into the ground. How much? That brings us back to the original question: How can you measure infiltration from a storm? You can't, at least for the present. No one has devised a way to measure infiltration accurately from precipitation. The best we can do is to estimate it from other kinds of measurements.

Although we can describe the physical process of water moving down through the surface into the soil, we cannot accurately measure either the rate of infiltration or the total amount infiltrated during a storm. We can determine the average depth of rainfall over a drainage basin, using one of the methods outlined in Chapter 4, and runoff, by measuring stream flow out of the basin. The difference between the rainfall coming in and runoff going out is considered the amount infiltrated during the storm. This obviously is not a precise measurement, since it ignores water losses such as evaporation, but it is often the closest approximation. At best, it is an estimate.

Another way to estimate infiltration for a drainage basin is to map all the soil types in the basin and make infiltrometer measurements on each soil type, similar to the Thiessen method for finding total precipitation over a basin. An *infiltrometer* isolates a given area of soil by placing a metal "fence" around the area reaching above and below the soil for several inches. Shoving a large can with both ends removed down into the soil would provide a very simple infiltrometer. The area of soil enclosed usually ranges from 1000 square centimeters (155 in.^{2}) to several square meters. Once the infiltrometer is in place, a sprinkler system that can simulate rain at a measured rate is installed. Water running off from the lower edge of the fenced area is caught and measured. The difference in water applied and water measured as runoff is assumed to be infiltration. Although this method can provide useful estimates, it tends to overestimate infiltration.

As with determining areal averages for precipitation, estimating infiltration requires considerable skill and experience from a hydrologist. Reality is not as simple as the bare outline presented here. Several factors must be considered besides total precipitation and runoff, for example: How wet was the soil (from previous rain) at the beginning of the storm? Were there variations in rainfall intensity during the storm? What is the surface texture of the soil? What about slope and vegetation in the basin? One obviously must have knowledge of local conditions and experience in evaluating the factors controlling infiltration in various parts of the drainage basin.

Regardless of how one arrives at a value for infiltration in a particular area, the process of infiltration itself is clearly an important part of the water cycle. When water falls on the ground, it either runs off in streams, goes up again as vapor, or sinks into the soil to become soil water.

SOIL WATER

Soil water is subsurface water in the unsaturated zone between the ground surface and the water table. It enters the soil through infiltration and either remains in storage or passes upward through evaporation and transpiration or it passes downward by gravity to the water table and the zone of saturation. During storms sometimes a component of subsurface water called *interflow* flows horizontally just beneath the surface. Even though it has passed through the soil surface as infiltration, interflow is not really soil water as defined here; Chapter 8 discusses it as part of the surface runoff.

Our interest in soil water as a part of the water cycle concerns its role as a supply for growing vegetation and as a source of recharging the underlying groundwater body. The unsaturated zone acts both as a storage reservoir and as a conduit for soil water in transit, and these two categories are convenient subheadings for a discussion of soil water. First, we need to examine the internal makeup of the soil mass to understand and discuss its storage and transmission functions.

Soil as a Porous Medium

The dictionary says a *pore* is a "small opening or passageway" and one definition of *medium* is an "environment." A porous medium is thus an environment full of small openings. Porous media take many forms: sponges, bread, the filter on the end of a cigarette, beds of sand and gravel, certain kinds of volcanic rock such as pumice, sedimentary rocks such as sandstone, soil, and so on. Some porous materials contain many small openings that are not connected to other openings. Pumice is one of these. These kinds of porous media are of little importance in the water cycle, because while they may be capable of storing a certain amount of water, they are incapable of transmitting water. Soil, on the other hand, is a medium capable not only of storing water but of passing it along through the water cycle.

All natural porous media containing interconnected pores share certain hydrological properties. These properties describe capacity for storing and yielding water; they are defined in terms of volume in underground space (see Figure 5.5). You will want to use Figure 5.5 as a reference when you encounter these terms in the pages ahead. Note that because the terms are volume ratios and therefore pure numbers, units are unimportant. For example, a *porosity* of 0.30 expresses the relative amount of open space in porous material, whether it is measured in cubic inches or cubic centimeters:

total soil volume: $V = 30\text{ in.}^3$ $\qquad V = 492\text{ cm}^3$

pore volume: $V_p = 9\text{ in.}^3$ $\qquad V_p = 148\text{ cm}^3$

porosity:

$$\frac{V_p}{V} = \frac{9\text{ in.}^3}{30\text{ in.}^3} = 0.30 \qquad \frac{V_p}{V} = \frac{148\text{ cm}^3}{492\text{ cm}^3} = 0.30$$

Porosity is often expressed as a percentage of total volume. Thus a porosity of 0.30 could also be expressed as 30% porosity. In other words, 30% of the soil's total volume is composed of open spaces between grains.

The porosity of unconsolidated granular material depends on the grains' shape, arrangement, and degree of assortment. Grain size is not important. All other things being equal, a porous material will have the same porosity whether it consists of large or small grains. A silty soil, consisting of very small grains, can have a porosity similar to that of coarse gravel. While the

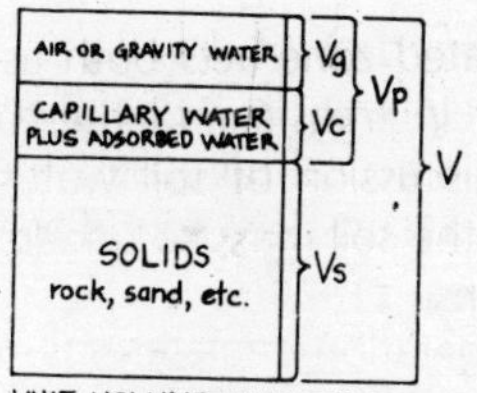

UNIT VOLUME IN UNDERGROUND SPACE

WHERE: V = TOTAL VOLUME
V_s = VOLUME OF SOLIDS
V_p = VOLUME OF PORES
V_c = VOLUME OF CAPILLARY WATER PLUS ADSORBED WATER
V_g = VOLUME OF AIR AND/OR GRAVITY WATER (PORE VOLUME MINUS CAPILLARY WATER)

THEREFORE: $V = V_s + V_p$
$V_p = V_c + V_g$

AND: POROSITY $= \frac{V_p}{V}$

FIELD CAPACITY $= \frac{V_c}{V}$ (APPROXIMATE)

SPECIFIC RETENTION $= \frac{V_c}{V}$

SPECIFIC YIELD $= \frac{V_g}{V}$

FIGURE 5.5 Some hydrologic properties of porous media.

porosity of silt and gravel may be equal, though, neither the size of pores in the two materials nor their other hydrologic properties will be the same. The proportion of pore space filled by capillary water will be much larger in the silt than in the gravel; hence, a silty soil will hold much more water in storage (against gravity drainage) than a gravel soil will. On the other hand, a gravel soil will be much more permeable than a silt, and gravity drainage will be faster and more complete from a gravel than from a silt.

Soil Water in Storage

The capillary water held by surface tension after gravity drainage has stopped is the maximum amount that the soil can hold in permanent storage. This is what soil scientists call *field capacity,* defined here as the ratio of the volume of capillary water per unit volume of soil, as shown in Figure 5.5. For a particular soil, this ratio represents a measurable amount of water per unit of depth. In books on agricultural and soil science, for example, field capacity is often defined as the *quantity* of water held against gravity. Thus for a given soil a field capacity of 23% could also be considered a capacity for holding 23 centimeters of capillary water per one meter of depth.

The terms *specific retention* and *specific yield* are used in groundwater hydrology and refer to the behavior of porous media below the water table in the saturated zone. Specific retention is almost the same as field capacity, except for the time element. Soil scientists usually consider field capacity to be capillary water held after a few days of drainage. Groundwater hydrologists usually consider specific retention to be capillary water held after a much longer drainage time. For practical purposes, the two are essentially the same, but some porous media have been found to yield a small amount of drainage water even after a year or more.

In the root zone, growing vegetation depletes soil water in storage. Osmotic pressure differences between fluid in the plant roots and water in soil pores allows the plants to extract water from the soil (see Chapter 6 for an explanation of osmosis). Most of the water drawn into plants is eventually discharged by transpiration from leaf surfaces; thus, whenever plants are active, soil water is continuously removed from storage. Even at night, when transpiration is nil, soil water may still be taken up to restore plant *turgor,* a measure of the fluid pressure within plant cells that gives herbaceous plants their rigidity.

As long as the suction force in the roots exceeds the capillary force in the soil pores, water flows from soil to plants. At some point, however, the pores' retentive forces may exceed the roots' ability to take up water. The residual amount (percentage) of soil water that cannot be extracted by roots is called the *permanent wilting point,* or wilting percentage of soil moisture for that particular soil. At this percentage of soil moisture, plants permanently wilt, and if no water is added to the soil, the plants will die. In the root zone, then, soil water in storage can vary from the percentage of water required for field capacity to the percentage of water still in storage at the wilting point. In a prolonged drought slow evaporation through the near-surface soil pores may reduce storage even further.

Below the root zone, evaporation to the surface is extremely slow, and once the full complement of capillary water is stored in the soil, it will remain almost indefinitely. Even in the driest deserts, soil a foot or so beneath the surface always contains at least a small percentage of moisture. This was discovered, for example, in the Colorado Desert in southeastern California during tests conducted on mineral concentrating equipment using dry, gravity separation of mineral grains to recover gold from a desert placer. Even in this very dry climate, excavated soil had to be dried before running it through the concentrator. In this desert, where rainfall is about 2 inches per year and evaporation rates are very high, the soil a foot or so below the surface still had a moisture content of at least 5%–6%.

Measuring Soil Water. Water content in the soil can be measured several ways. The simplest and most direct, called the gravimetric method, is to take a sample of moist soil, weigh it, dry it in an oven at 105° C for 24 hours, weigh it again to determine the water loss, then express the moisture content as a percentage of the dry weight. To be consistent with the concepts of volume ratios discussed previously, moisture content should be expressed as a volumetric percentage rather than as a weight percentage. This would involve determining the soil's bulk density, or dry weight per unit volume. It is done by weight simply for convenience—accurate volume determinations on small, disturbed samples of soil are much more difficult than accurate weight determinations.

A disadvantage of the weighing and drying method is that the sample is taken from the soil and cannot be replaced as it was before. Future measurements cannot be made at that precise location. When you want to monitor changes in water storage in the soil profile, you need a method that can measure moisture content without disturbing the surrounding soil. Several methods can accomplish this; two that are used extensively are the *tensiometer* and the *neutron probe*.

A tensiometer measures soil-water "tension" or "suction." In other words, it measures the force with which soil pores hold capillary water. This is not a direct measurement of the amount of soil water in place. When the relationship between soil-water tension and water in a particular soil is known, however, readings from a tensiometer sometimes can be converted to approximate the soil's water content. This method of determining actual water content may be more or less accurate depending on whether the soil is undergoing wetting or drying at the time of measurement. Also, while tensiometers are reliable tools for use in moist soils, they do not work well in dry soils.

A common type of tensiometer consists of a sealed, water-filled tube with a vacuum gage at the upper end and a porous ceramic tip that goes into the ground at the lower end (Figure 5.6). Water in the tube is in hydraulic continuity with soil water through the porous tip. As the soil dries, the soil-water tension increases and capillary forces in the soil suck water from the tube, creating a partial vacuum, which the gage at the above-ground end of the tube detects. Adding water (to soil around the tip) through natural rainfall or applied irrigation will reduce soil-water tension and consequently lower the reading on the vacuum gage. The instrument will record a maximum vacuum of about 0.85 atmospheres. The vacuum gage can be calibrated to read in atmospheres, pounds per square inch, kilopascals, or any convenient measure of fluid pressure.[3] In practice, gage calibrations are often interpreted in purely relative terms, indicating soil conditions to be wet, moist, or dry. Tensiometers are used mostly by farmers to help in scheduling irrigation of growing crops. The instruments come in standard lengths from 6 inches to 60 inches (15–152 cm).

Tensiometers are relatively inexpensive, useful tools for the farmer as long as their limitations are known. Farmers want to know not only how much water is in the soil, but how available it is for extraction by crop roots. As

[3]An *atmosphere* is a unit of pressure equivalent to the weight of the earth's atmosphere at sea level at 0° C. One atmosphere equals 14.7 lb/in.^{2} or 101.3 kPa.

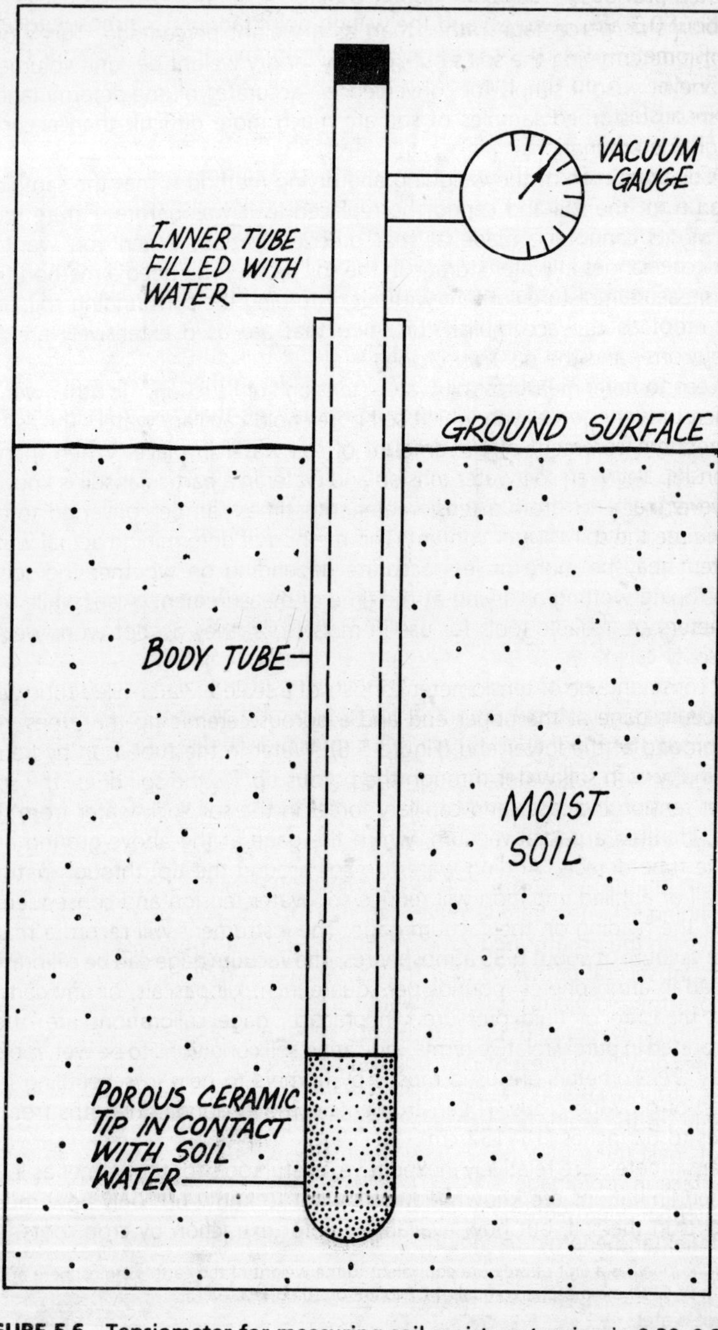

FIGURE 5.6 Tensiometer for measuring soil-moisture tension in 0.30–0.85 atmosphere range.

stated previously, roots can extract water in a range between field capacity (about 0.3 atmospheres) and the wilting point (about 15 atmospheres). The tensiometer measures soil suction only from about 0.3–0.85 atmospheres, however. To do this the porous tip must be deep enough to touch moist soil constantly.

The neutron probe is an atomic-age instrument that has proven useful in measuring water in place at any depth and in any quantity. For neutron-probe measurements, small drill holes, cased with 2-inch metal or plastic tubing, provide access for repeated, nondestructive measurements over a long period. Aluminum is the preferred material for casing the holes because it is essentially transparent to neutrons. The casing tubes usually extend below the root zone. The probe has a radioactive source that emits high-energy neutrons at a nearly constant rate, and these in turn bombard the soil adjacent to the drill hole (Figure 5.7). A recorder in the tool measures the rate of returning low-energy (epithermal) neutrons that result from the bombardment. When the high-energy neutrons enter the soil, they collide with certain atoms in the soil matrix and are reduced in energy, returning as epithermal neutrons to be counted by the measuring tool. The chief energy reducer is the hydrogen atom, and the tool is therefore hydrogen-sensitive. Since nearly all of the hydrogen in soil occurs in water, the tool measures the water content. In saturated soils, water content is a function of the amount of pore space. For soil-water measurements, the surface recorder is usually calibrated to yield volume percentage of soil moisture. For the soil scientist, the neutron probe gives valuable information on changes in soil moisture as plants extract water for growth and as water is replenished and drained after rainfall or irrigation. Chapter 6 discusses use under actual field conditions in California.

Soil Water in Motion

As in infiltration, gravity and forces associated with the action of hydrogen bonding in water, here termed "capillary" forces, are responsible for moving water in the unsaturated zone. A drop of water infiltrating the soil may take one of several pathways as it moves through the unsaturated zone. It may move toward plant roots, or toward the soil surface (where it will be transpired or evaporated and returned to the atmosphere). Once past the root zone, gravity may eventually pull it down to the water table and it will join the groundwater in the saturated zone. On occasion the drop of water may infiltrate the soil in a farming area containing subsurface drains. Here, if it gets past the root zone, a drain line may intercept it and discharge it at the surface in a drainage ditch. Water will always be controlled by the existing potential-energy gradient—that is, flow is always from regions of high potential (and energy) to regions of lower potential.

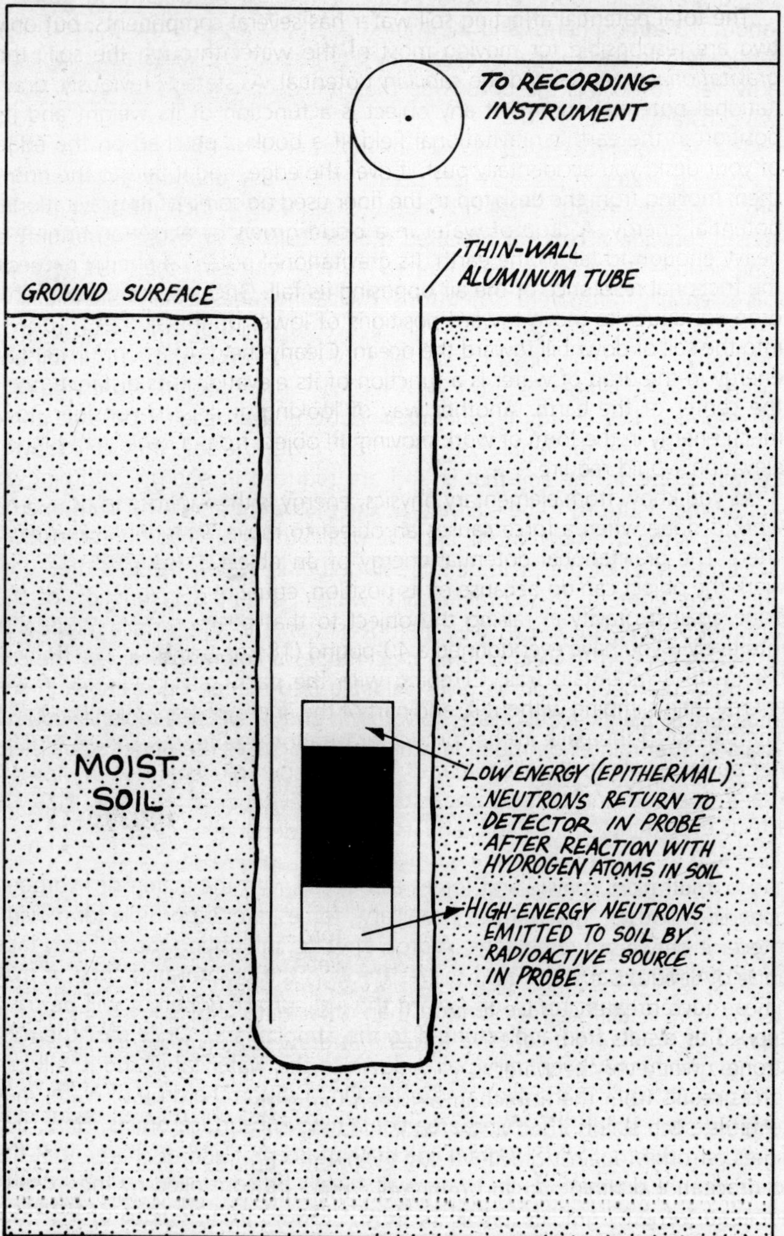

FIGURE 5.7 Neutron soil probe.

The total potential affecting soil water has several components, but only two are responsible for moving most of the water through the soil: the *gravitational* potential and the *capillary* potential. As stated previously, gravitational potential energy of any object is a function of its weight and its position in the earth's gravitational field. If a book is perched on the edge of your desk, you accidentally push it over the edge, and it falls to the floor, then, moving from the desk top to the floor used up some of its gravitational potential energy. A drop of water in a cloud grows by accretion until it is heavy enough to fall to the earth. Its gravitational potential energy exceeds the frictional resistance of the air opposing its fall. Once on the ground, the drop continues to move toward positions of lower potential energy, down into the soil or downhill toward the ocean. Clearly the gravitational potential energy in the drop of water is a function of its elevation—its distance from the center of the earth. Another way of looking at it is to consider how much energy in the form of work moving an object from a lower to a higher elevation would require.

As you know from elementary physics, energy is the capacity to do work; work is done when a force causes an object to move, as when a weight is lifted. The gravitational potential energy of an object is a measure of the work the object can do because of its position, equal to the amount of work done against gravity in raising the object to that position. You know this intuitively if you have ever donned a 40-pound (18-kg) backpack and started hiking up a mountain trail. Climbing with the pack definitely takes more energy than climbing without it, and part of the energy expended in climbing is imparted to the pack in the form of gravitational potential energy.

Capillary potential (called *matric potential* by soil scientists) is not as straightforward and easy to understand as gravitational potential. Gravity affects everything on the earth, and its strength depends only on the object's mass and its distance from the earth's center. Capillary forces, on the other hand, result from cohesive and adhesive molecular forces acting within the body of a liquid. Cohesive forces cause surface tension, which acts at the interface between two liquids or between a liquid surface and a gas (such as air). All liquids exhibit surface tension, but as Chapter 2 showed, water has a much greater surface tension than other natural liquids (except mercury). This results from cohesive forces that draw water molecules together strongly wherever the water is in contact with air or other gases. Adhesive forces result from the attraction between water molecules and most solid surfaces, such as adsorbed water, which is a thin film on a solid surface that is held strongly against gravity. It will evaporate, but it will not run off due to gravitational force. Do you remember the cause of these cohesive and adhesive forces? (Hint: its initials are H.B.).

The forces that cause water to rise in a capillary tube are always present. If you pour water into a drinking glass and examine its sides, you can see the water surface curve up slightly where it is in contact with the glass. Water molecules have been adsorbed on the glass (the glass has become "wet"), and the molecules at the water surface are strongly attracted to each other, tending to reduce the surface area in contact with the air. While this is barely noticeable in a vessel as large as a drinking glass, if you were to shrink it to a centimeter or less in diameter you would see a steady rise in the water surface. Molecular forces within the water "reach" up to wet the glass at a higher point while the forces of surface tension act to reduce the liquid's surface area. This results in a "climbing" action as the water rises, with the molecular forces increasing as the water-air surface area decreases—that is, as the glass tube diameter shrinks. Thus capillarity, which depends on molecular forces of adhesion with a solid surface and surface-tension forces within the liquid, raises water in opposition to the downward force of gravity. In an extremely small tube or succession of soil pores, capillary rise may move water many feet above an open water surface such as a water table. Capillarity, however, always operates in the earth's gravitational field, so the rise of water in a tube or soil pore will continue only until the gravitational force balances it.

Capillary potential energy, then, depends on molecular forces within a liquid and on tube or pore diameter. This explains why when two soils with different pore sizes such as sand and silt are in contact underground, the capillary potential-energy gradient will usually run from the sand toward the silt (i.e., from larger to smaller pores). The capillary potential-energy gradient can take any direction in the soil—up, down, or horizontal—with water moving from wetter toward drier soil.

Soil water will tend to move in response to the total potential-energy gradient, which in turn results mainly from a resolution of gravitational and capillary forces. Thus rainfall is literally "sucked" down into dry soil. Water moves horizontally in soil capillaries toward dry soil around plant roots, or upward toward the soil surface as the ground begins to dry after a rain. When soil is saturated, menisci are destroyed and the capillarity disappears temporarily as gravity takes over and water drains downward. As the supply diminishes and drainage removes more water, menisci begin to reappear and capillarity again takes over, holding water in the pores.

Actual conditions in nature are usually more complex than those described here. Air in pores may, and often does, prevent complete saturation during infiltration and gravity drainage, and that complicates flow conditions. A more complete explanation of permeability and flow through porous media will be given in Chapter 7 when flow in the saturated zone is discussed. The description of flow in the unsaturated zone given here is oversimplified and is of value mainly in understanding the conceptual basis of soil water movement. Some of the books listed in Appendix IV contain more technical descriptions of soil water movement.

The zone of soil water (the unsaturated zone) is like a vast sponge that soaks up water from precipitation, holds it a while, then passes it on to groundwater flow or to evaporation and transpiration. Its function under the ground parallels that of the atmosphere above ground. As does the atmosphere, the unsaturated zone holds water in transit for varying periods. Eventually, though, most of that water returns to the surface and continues on its inevitable journey back to its origin in the oceans.

APPLICATIONS

Infiltration applications focus mainly on increasing the natural supply of underground water, both for use by plants and for recharge of the underlying groundwater reservoir.

Irrigation

Irrigation is the chief means for increasing the water supply to plant roots. Irrigation agriculture has long supported human habitation in arid regions. Until recent years, methods for applying irrigation water to land had changed little since the Egyptians invented this method of farming more than 5,000 years ago.

Traditionally, irrigation has allowed water to flow into shallow basins where crops have been planted or has directed it down furrows between rows of plants. Either method meant applying water in substantial quantity directly to the soil surface. Near the turn of the century, U.S. farmers added a new dimension to irrigation by using overhead sprinklers to water crops. The first major innovation in irrigation agriculture in several millennia was gradually adopted by farmers in other countries, and today sprinklers water crops in many farming regions around the world.

Sprinklers have not displaced basin and furrow irrigation, but they do have distinct advantages in many instances, including better erosion control, especially on land with a steep slope; better water distribution on sandy soils with high infiltration rates; conservation of water, allowing use of a smaller natural flow from a stream or canal to irrigate more land than by traditional basin and furrow methods; and other advantages, not the least of which is generally reduced labor required for irrigating the fields. Even though sprinklers often use less water to irrigate a crop than basin or furrow methods would require, they still use a substantial amount of water, some of which is lost to drift in the wind, evaporation in the air, interception by plant surfaces, and evaporation from wet soil surfaces In arid regions where water is really in short supply, even sprinklers are sometimes not efficient enough for widespread use.

A truly revolutionary method of applying water to the soil has recently been developed for farmers with access to a limited water supply. The new method, called "drip" or "trickle" irrigation, is not really all that new—greenhouses have used it on a limited scale since the early days of this century for irrigating plants. Full-scale development on a commercial basis for trees and field crops has occurred only within the last 20 years, though. Most early experiments with the method took place in Israel, followed a few years later by widespread application in California. Although farmers in Europe and a few other regions have actively applied drip irrigation to field practice, the following discussion is based mostly on experience in Israel and in California.

Drip Irrigation. Before describing the method, it should be noted that drip irrigation has some of the same advantages as sprinklers, without most of the water-loss disadvantages. As will be evident in the discussion that follows, however, the increased irrigation efficiency of the drip method may also mean higher costs, sometimes much higher costs. It is primarily a technique for regions with an absolute limit to the available water supply or where the water quality is marginal for other methods but still acceptable for drip or trickle methods.

Drip irrigation applies water through small emitters to the soil around individual plants or trees, as shown in Figures 5.8 and 5.9 (for location, see Map in Appendix I). A system of plastic pipes and hoses, operating at low pressure, delivers water to the emitters at rates of about 1–3 gallons (4–11 L) per hour. Water released from the emitters wets the surface and is drawn down by capillary suction and gravity into the soil, where it becomes available for uptake by roots and eventual transpiration by the plant.

Some of the more important advantages in using drip irrigation include:[4]

1. *Water conservation.* Some crops can grow with much less water under drip as compared with other irrigation methods. Young orchards (see Figure 5.8) may require only about half as much water by drip as by sprinkle or basin irrigation. The advantage is somewhat less in mature orchards, but it is still significant where water is scarce.
2. *Less evaporation from soil.* Because only a small area of soil is watered, losses through evaporation from wet soil are greatly reduced.
3. *Lower labor costs.* When properly designed and operated with automatic controls, drip systems can provide considerable savings in labor costs. Because most of the soil surface remains dry, weeds are not a serious problem, so weed control is unnecessary. Labor for applying fertilizers can be almost entirely eliminated by injecting fertilizer into the irrigation water. Improved efficiency in using fertilizers

[4]This section on advantages and disadvantages is adapted mainly from the chapter on trickle irrigation in the U.S. Soil Conservation Service National Engineering Handbook (1980).

FIGURE 5.8 Drip irrigation in a young almond orchard on the west side of the San Joaquin Valley near Mendota, California. Water is delivered to each tree by a buried plastic pipe. An emitter releases 1.5 gallons (5.7 L) per hour to soil at base of tree (photo by U.S. Bureau of Reclamation).

is also possible, since the emitters can control placement and timing of water deliveries.

4. *Water quality less important.* Because frequent applications from the drip emitters can maintain soil moisture at more desirable levels, this method may allow use of more saline water than with any of the more traditional methods.
5. *Works well on any slope.* Slope steepness is unimportant, so land does not need to be leveled for efficient use of the method. In some avocado orchards under drip irrigation in San Diego County, California, slopes are so steep that from a distance the trees appear to be growing on the face of a cliff. Harvesting the fruit here might be more of a problem than irrigating and fertilizing the trees.

In spite of the drip method's considerable advantages, it also has some serious disadvantages:

FIGURE 5.9 Drip irrigation in a mature almond orchard on the west side of the San Joaquin Valley near Mendota, California. The drip hose, fed by a buried plastic pipeline, has four emitters, each of which is releasing 1.5 gallons (5.7 L) per hour to the soil around base of tree. The emitters are at the centers of the four wet spots seen in the photo (photo by U.S. Bureau of Reclamation).

1. *Cost.* A chief disadvantage is the high initial cost for a properly designed system with all the necessary automatic controls and water-treatment and filtration equipment. A poorly designed system may cost so much to operate that it is eventually abandoned. Two problems that the design must address are clogging and lack of uniform rates throughout the water-delivery network.
2. *Clogging.* The holes through which the emitters release water to the soil are small, ranging from a minimum diameter of 0.008 inches (0.20 mm) to 0.060 inches (1.5 mm) and larger. Irrigation water often contains particles of mineral and organic matter large enough to clog these tiny openings. In some waters, mineral precipitates tend to form in the lines or at the emitter openings. For a drip system to function properly, then, the water must be absolutely clean and chemically stable. This almost always means passing the water through a fine-

mesh filter to remove all particulate matter. Depending on the quality of the local supply, it may also mean extensive chemical treatment to prevent formation of mineral precipitates downstream from the filters. As you might imagine, filtration and treatment can substantially increase overall capital costs and operating expenses.

3. *Uniform flow.* This is a problem in a system installed on steep slopes. As most drip emitters operate at low pressures (3–20 pounds per square inch, or 21–138 kilopascals), discharge from emitters at the top and bottom of a steeply sloping field may vary as much as 50% from the required amount. Water may also drain from lower emitters after the supply is shut off at the end of an irrigation cycle. Thus, some plants receive too much water and others too little.
4. *Unfavorable soil conditions.* Even in a well-designed system, soil conditions may prevent efficient operation. In that respect drip irrigation is no different than any other method; no irrigation will work if water doesn't penetrate the ground surface. If the flow from the emitters exceeds the soil's infiltration capacity, water will collect in ponds and will begin to flow downslope as surface runoff. Fine-grained soils, especially those with deflocculated clay particles on the surface, may not be responsive to drip irrigation, and applying soil amendments such as gypsum to reflocculate the clays probably is not practical in most areas using drip irrigation. Medium- to coarse-grained sandy soils are the most suitable for drip irrigation. Many of these soils, especially valley and floodplain soils, originated from sediments transported by wind or water and hence have some horizontal stratification. This tends to cause infiltrated soil water to spread laterally and wet a larger volume of soil than would an unlayered, homogeneous body of soil.
5. *Salt deposits.* Drip irrigation has been used mostly under arid or semiarid conditions where high evaporation opportunity may cause salt deposits to accumulate on or just beneath the surface of the ground. Salts below ground are apt to concentrate around the perimeter of moist soil surrounding the plant roots, and this salt may tend to move inward toward the roots as the soil dries between irrigations or at the end of the irrigation season. Salt movement could also occur if the soil dried during a system failure when the emitters stopped delivering water. Regardless of how it occurs, the salt must not be allowed to move close to the plant roots. Depending on the rate and total volume of salt accumulation, most cases will require putting enough water on the ground to leach the salts down into the soil below the root zone. Water to flush out the salt could come from

winter rains or from excess irrigation water applied at the end of the irrigation season.

Whatever its advantages and disadvantages, drip irrigation is firmly established as an alternative method of irrigation where environmental and economic conditions favor its use. It is offered here as an example of applying theoretical knowledge of infiltration and soil water to solve a practical problem; namely, how to grow crops where even marginal quality water is in short supply.

Groundwater Recharge

Another example of practical applications for infiltration is the operation of infiltration ponds for recharging the underlying groundwater reservoir. Recharge ponds demonstrate almost the reverse principle of drip irrigation. The drip method essentially maintains a supply of soil water at or near field capacity in the root zone to nourish growing plants. It allows deep percolation by gravity drainage below the root zone, only for leaching salts from this zone. Drip emitters discharge as little water as possible. Groundwater recharge, on the other hand, infuses all the water into the ground that the soil will accept, allowing the maximum amount to pass downward by deep percolation to the water table. Both techniques aim to conserve water, but drip irrigation immediately conserves a limited supply, while groundwater recharge conserves an excess supply now for use in the future.

Water for groundwater recharge is commonly excess runoff water that normally cannot be stored on the surface until it is needed. This might be floodwater moving downstream after a big storm, or water released from a reservoir of runoff stored from past storms.

Using floodwater for recharge serves a dual purpose, diverting stream flow into off-channel spreading basins and helping to reduce flood crests downstream. When floodwater is used, however, most of the sediment must be removed from the water before diverting the flow into the spreading basins. Otherwise, the basins would soon fill with sand and silt, as in the case of the sand-storage dams in Namibia described in Chapter 3.

One of the most extensive series of off-channel spreading basins for recharging floodwater was developed by various public flood-control and water-conservation districts on the coastal plain of southern California. Here, large debris dams built at the mouths of mountain canyons can catch most of the sediment brought down the canyons by flood flows. While they operate on a much larger scale than the Namibian sand-storage dams, the debris dams are just as effective in catching most of the sediment carried by floodwater. The spreading basins for groundwater recharge are adjacent to the flood channels downstream from the debris dams. Here, floodwater, relatively free of suspended sediment, can infiltrate the underground to recharge the extensive groundwater reservoirs supplying water to the coastal plain in Los Angeles and Orange Counties. Unlike in the Namibian sand-storage dams, sediment in the southern California debris dams is periodically removed and disposed of to maintain space for sediment brought down by future floods.

Another region of California that uses infiltration ponds extensively for groundwater recharge is the San Joaquin Valley. All the major streams flowing from the Sierra Nevada, which forms the eastern boundary of the valley, have dams and reservoirs to store the spring runoff from the Sierran snowpack. Much of this stored water is released to a large network of canals that carry irrigation water to farms on the valley floor. In years of heavy precipitation, however, the reservoirs won't hold all the runoff and the excess water either must be allowed to flow into San Francisco Bay and the ocean or it must be stored in that largest and most extensive of all reservoirs, the groundwater reservoir. To do this, water is released from the surface reservoirs during the spring runoff and put into infiltration ponds for ultimate transfer to the underground.

Arvin-Edison Water Storage District. One of the most successful recharge projects in California is operated by the Arvin-Edison Water Storage District. The district, located southeast of Bakersfield along the eastern edge of the San Joaquin Valley (see map in Appendix I), contains about 112,000 acres (45,326 ha) of irrigated land. The major crops are cotton, grapes, potatoes, citrus, almonds, and a variety of vegetables. With no perennial streams in the district, and only about 8.5 inches (216 mm) of rain a year, farmers must irrigate their crops. Before the advent of imported surface water in 1966, farmers depended entirely on groundwater pumped from their own wells. Over a number of years, that pumping had drastically reduced groundwater levels. To augment the well-water supply and stabilize groundwater levels, the land owners formed the Arvin-Edison Water Storage District, which in turn contracted with the U.S. Bureau of Reclamation to supply supplemental irrigation water. The water comes by canal from Friant Dam on the upper San Joaquin River east of Fresno, about 150 miles (240 km) north of the district. There are two classes of water, Class 1 (firm water) and Class 2 (nonfirm water). Class 1 water comes mostly during the irrigation season and is usually applied to the land as soon as it arrives in the district. Class 2 water usually comes in the spring months and is used both to irrigate and to recharge groundwater through an extensive network of infiltration ponds. The following discussion is limited to a description of the ponds and how they work. What happens to the water once it is in the ground and what this has meant in terms of the district's ability to serve the farmers is discussed in detail at the end of Chapter 7.

Infiltration Ponds. The infiltration ponds of the Arvin-Edison WSD fall into two areas under which permeable alluvial sediments lie: one along Sycamore Creek near the middle of the district, and the other about 6 miles (10 km) to the south along Tejon Creek. Both creeks are ephemeral streams that carry seasonal runoff from the mountains bordering the district on the east. The Sycamore ponds cover an area of 390 acres (158 ha) and the Tejon ponds, 516 acres (209 ha). Figure 5.10 shows an aerial view of the Sycamore ponds downslope from the main canal that carries surface water through the district. Gravity carries water from the canal into the Sycamore ponds and into about one-half of the Tejon ponds. Pumps supply the upslope Tejon ponds, which are used only when the water supply exceeds the capacity of the downslope ponds.

In common with most operators of infiltration systems, the Arvin-Edison WSD has found the main operational problem to be maintaining satisfactory infiltration rates over time. When dry ponds are first filled, it is not unusual for the water to disappear rapidly for a while; with time, the infiltration rate typically declines slowly until it reaches some lower limit at which it may

FIGURE 5.10 **Sycamore infiltration ponds used for recharging groundwater beneath the Arvin-Edison Water Storage District in the southeastern San Joaquin Valley, California. View is from the air looking northwest (photo by C. E. Trotter).**

then stabilize. The declining infiltration rates are caused by restrictions to the water's flow path either at the soil surface or at some choke-point in the soil pores between the surface and the water table. Little can be done about a subsurface restriction to flow, but some things can be done at the soil surface.

Soil pores may be clogged by fine particulate matter carried in water just as emitters in a drip-irrigation system are. Considering the much larger volume of water going into infiltration ponds as compared with drip emitters, mechanically filtering the water as is done in drip systems would not be practical. A more practical solution would be to monitor the water quality and try to prevent really turbid water from going into the ponds. At Arvin-Edison, for example, when suspended solids in the water supply exceed 25 parts per million, the water is no longer put into the ponds; however, if over time a substantial volume of fine silt does get into the ponds, the ponds are dried out periodically and the silt is removed mechanically.

Letting the soil surface dry periodically seems to help increase infiltration rates regardless of whether a lot of silt has accumulated. Planting grasses and developing a turf on the soil surface also aids in restoring or maintaining favorable infiltration rates. Periodically scarifying the surface by discing or otherwise stirring or loosening the soil seems to help. Finally, some operators of infiltration ponds have found it helpful to grow a crop and leave some of the crop residue on the surface or disc it into the soil. In other cases where no crop is grown, operators in the San Joaquin Valley have disced cotton-gin trash into the soil. This coarse, organic residue is obtained from one of the many cotton gins operating in the valley.

Regardless of the method, the Arvin-Edison operators have managed to put a lot of water underground. Out of about 2.6 million acre-feet (3,207,100,000 m^3) of water imported since 1966, approximately 800,000 acre-feet (986,800,000 m^3) have been put underground through the infiltration ponds. The remainder was delivered directly to the farmers for use on the land. Experience at Arvin-Edison has shown how successful a joint surface-water ground-water project can be when it is operated efficiently. The full story, detailing how stored water was pumped out of the ground to help farmers during a serious drought, is told in the Application section at the end of Chapter 7.

6

Evapotranspiration

The thirsty earth soaks up the rain,
And drinks, and gapes for drink again.
The plants suck in the earth, and are
With constant drinking fresh and fair.

(A. Cowley)

If you understand these lines from Abraham Cowley's poem, "Anacreon," you already know what this chapter is about. Three hundred years before the term *evapotranspiration* was invented, Cowley was able to summarize the process succinctly and clearly. We can hardly improve on his description. All that remains now is to learn something of the mechanics of the process. The last chapter described how "(t)he thirsty earth soaks up the rain." This chapter explores how "(t)he plants . . . are with constant drinking fresh and fair."

Evapotranspiration refers to the three major processes that return moisture to the atmosphere over the vegetated land areas of the earth, including:

1. Evaporation of precipitation intercepted by plant surfaces.
2. Evaporation of moisture from plants through transpiration.
3. Evaporation of moisture from the soil surface.

Because attempting to separate the individual effects of these processes is impractical, they are lumped together as the total process of *evapotranspiration.* An equivalent term also commonly used to describe total evaporation from a vegetated area is *consumptive use.* Consumptive use differs from evapotranspiration only in that it includes the water used to make plant tissue. For all practical purposes these two terms are synonymous; this chapter will use them interchangeably.

Globally, evapotransportation involves a substantial amount of water, giving it an important role in the water cycle. Long-term studies indicate that over large land areas in the temperate zones, evapotranspiration returns about two-thirds of the annual precipitation to the atmosphere, with only about one-third running off in streams to the ocean. In arid regions, evapotranspiration may be even more significant, returning up to 90% or more of the annual precipitation to the atmosphere. In the tropics, particularly with monsoonal rainfall, the proportion of runoff probably is somewhat larger and evapotranspiration less than in temperate regions.

Earlier chapters have discussed evaporation at length, and the evaporation of soil moisture or moisture intercepted by vegetation during a storm is subject to the same conditions as evaporation from an open water surface. Transpiration, however, whereby water evaporates from plant cells, is a different kind of process. In contrast to open water or moist soil, plants have some control over when and to what extent they discharge moisture to the atmosphere.

THE PROCESS OF TRANSPIRATION

Have you ever walked into a greenhouse and suddenly felt the increase in humidity? You probably realized then that plants were adding moisture to the air. Another time, perhaps on a hot summer day, you may have walked or driven from open sunshine into the deep shade of a forest glade and felt sudden cooling in the air. Again the air was more humid, and part of the cooling came from the large amount of heat energy removed from the air during evaporation of moisture from plant surfaces. *Transpiration,* the evaporation of water that has passed through plants, caused both phenomena (see Figure 6.1).

Transpiration occurs at all exposed parts of plants, but by far mostly in tiny openings in the leaves called *stomata* (the singular is *stoma* and comes from the Greek word, meaning mouth). Stomata are very small. Anywhere from 50 to 500 may cover a square millimeter of leaf surface, with many more on the lower leaf surface than on the upper. Stomata tend to open and close as the plant responds to environmental conditions such as light and dark, heat and cold, and so on. Their function is vital to plant metabolism because these openings allow carbon dioxide to enter the plant in the process of photosynthesis. The opening and closing of stomata affect the rate of transpiration and consequently control the flow of water upward from the

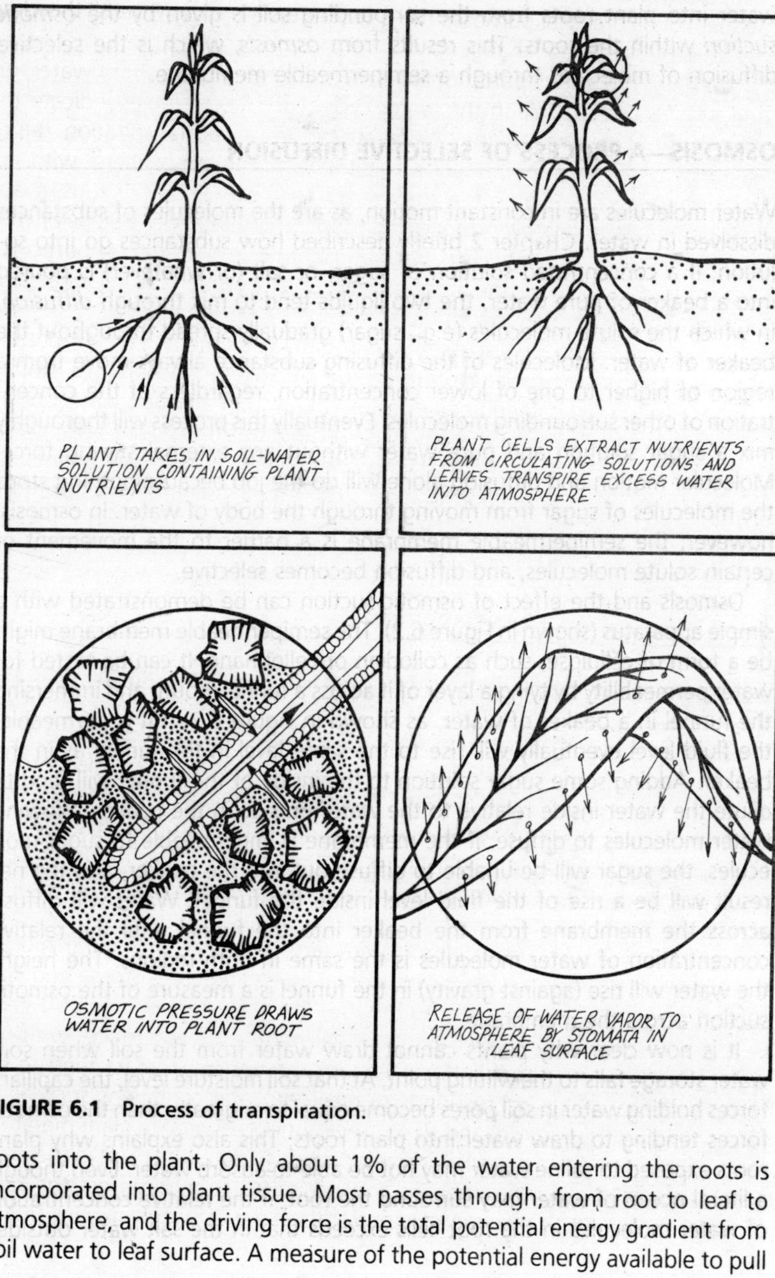

FIGURE 6.1 Process of transpiration.

roots into the plant. Only about 1% of the water entering the roots is incorporated into plant tissue. Most passes through, from root to leaf to atmosphere, and the driving force is the total potential energy gradient from soil water to leaf surface. A measure of the potential energy available to pull water into plant roots from the surrounding soil is given by the *osmotic suction* within the roots. This results from *osmosis,* which is the selective diffusion of molecules through a semipermeable membrane.

OSMOSIS—A PROCESS OF SELECTIVE DIFFUSION

Water molecules are in constant motion, as are the molecules of substances dissolved in water. Chapter 2 briefly described how substances go into solution. If a concentrated solution of sugar, or salt (or whatever) is poured into a beaker of pure water, the two liquids tend to mix through *diffusion,* in which the solute molecules (e.g., sugar) gradually spread throughout the beaker of water. Molecules of the diffusing substance always move from a region of higher to one of lower concentration, regardless of the concentration of other surrounding molecules. Eventually this process will thoroughly mix a sugar solution and pure water without any external stirring force. Molecular motion and diffusion alone will do the job because nothing stops the molecules of sugar from moving through the body of water. In osmosis, however, the semipermeable membrane is a barrier to the movement of certain solute molecules, and diffusion becomes selective.

Osmosis and the effect of osmotic suction can be demonstrated with a simple apparatus (shown in Figure 6.2). The semipermeable membrane might be a form of cellulose, such as collodian or cellophane. It can be tested for water permeability by tying a layer of it across a funnel mouth and immersing the funnel in a beaker of water, as shown in Figure 6.2. If it is permeable, the fluid level eventually will rise to the same level in the funnel as in the beaker. Adding some sugar solution to the inside of the funnel will tend to dilute the water inside relative to the water outside in the beaker, allowing water molecules to diffuse. If the membrane is impermeable to sugar molecules, the sugar will be unable to diffuse out into the beaker, and the net result will be a rise of the fluid level inside the funnel. Water will diffuse across the membrane from the beaker into the funnel until the relative concentration of water molecules is the same in both regions. The height the water will rise (against gravity) in the funnel is a measure of the osmotic suction across the membrane.

It is now clear why plants cannot draw water from the soil when soil-water storage falls to the wilting point. At that soil moisture level, the capillary forces holding water in soil pores become equal to or greater than the osmotic forces tending to draw water into plant roots. This also explains why plant roots exposed to saline water may not be able to absorb water. Even though a literal ocean of water may surround the root, if the relative concentration of water molecules in the root cells exceeds that in the salt water outside,

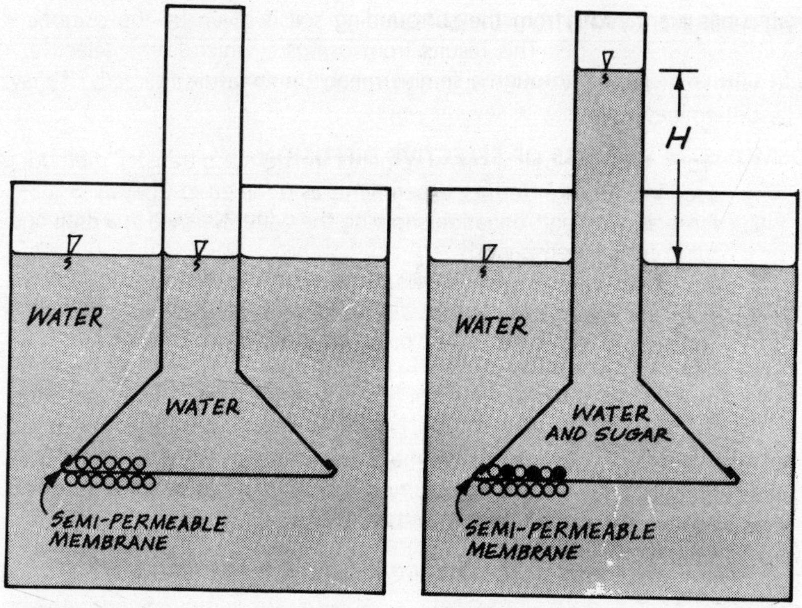

FIGURE 6.2 Demonstration of osmosis. The height (H) is a measure of the osmotic suction across the semipermeable membrane.

selective diffusion will tend to cause water to move from the roots into the soil. A small group of salt-resistant plants, called *halophytes,* are able to take in water from saline soils by greatly reducing their rate of transpiration, thereby effectively reducing the concentration of water molecules and increasing the force of osmotic suction in their root cells. For most plants, however, soil containing salty water is said to be physiologically dry, even when it is saturated.

The semipermeable membrane in plant roots is made of protoplasm and selective diffusion is much more complicated than in the simple example shown in Figure 6.2. Osmosis transports both water and certain inorganic nutrient molecules into the roots and thence to all living cells of the plant. The constant discharge of water from the plant through transpiration keeps the relative concentration of water in the roots lower than that in the soil water and permits osmosis to occur. The complete process of water flow from roots to leaves to atmosphere is complex and still not completely understood. It is clear, however, that in the context of the water cycle, plant transpiration is a major factor in transferring water temporarily stored in the soil back intc the atmosphere.

MEASURING EVAPOTRANSPIRATION

As with infiltration, evapotranspiration cannot be measured directly. It may be determined indirectly in several ways:

1. From meteorological measurements in the mass-transfer method, as used in the Lake Heffner experiment, as outlined in Chapter 3.
2. From a hydrologic equation showing the water balance in a drainage basin for a given period.
3. From measurements made with plants grown in lysimeters (soil tanks).
4. From soil-water-depletion studies using methods such as gravimetric sampling or the neutron soil probe, as outlined in Chapter 5.

Basin Water Balance

In a drainage basin, the water cycle may be described in terms of water input and output. In a humid region, assuming precipitation to be the only input, a water-balance equation can be written as:

$$\text{Evapotranspiration} = \text{Precipitation} - \text{Runoff} \pm \text{Changes in Storage}$$

Storage here refers to both soil water and groundwater as well as to surface lakes and reservoirs. Precipitation and runoff are both measured over a period long enough (a year or period of years) that the net change in water storage is essentially zero within the basin. Therefore, the difference between long-term measurements of precipitation and runoff is estimated to be the quantity of evapotranspiration for that time period. Computed in this way the estimated basin evapotranspiration includes evaporation from lakes and reservoirs. However, these quantities can be estimated separately by Class A pan measurements (discussed in Chapter 3).

Precipitation may not be the only input to a basin in an arid or semiarid region; in addition, groundwater storage may change considerably over time. For example, in the southern San Joaquin Valley in California (see map in Appendix I), inputs include not only precipitation but inflow from canals and surface streams as well. Because this is essentially a closed basin, in most years there is no surface runoff out of the basin. Almost all the water output is through evapotranspiration from about 3.3 million acres (1,335,510 ha) of irrigated land. For this semiarid basin the water-balance equation could be written as

$$\begin{aligned}\text{Evapotranspiration} = {} & \text{Precipitation} + \text{Input from Stream and} \\ & \text{Canal flow} \pm \text{Changes in Storage}\end{aligned}$$

"Changes in storage" refer mainly to groundwater. Because irrigation demand has exceeded surface-water supplies in most years during the past four decades, groundwater storage has steadily declined. This is explained in detail in Chapter 7.

Lysimeter Method

A lysimeter is a tank buried in the ground, its top just level with the surrounding soil surface, filled with soil similar to that surrounding the tank. It is an instrument for measuring the water balance in the contained soil. Lysimeters have been used for many years in research on evapotranspiration of particular plant species, especially those grown as cultivated crops. Although both weighing and nonweighing lysimeters are in common use, the weighing type is probably more useful for studies of evapotranspiration during periods shorter than a growing season. Both types require drainage at the bottom (see Figure 6.3). All other aspects of the soil and vegetation should be as similar as possible to the soil and vegetation surrounding the instrument.

Either type of tank should include some provision for measuring the change in soil-water storage. Furthermore, once soil is in place and plants begin to grow, the soil structure should not be disturbed. In the nonweighing lysimeter, irrigation water added must be carefully measured. A neutron probe can periodically measure resulting changes in soil water. For the weighing type, changes in weight indicate changes in soil water; scales measure these. These values can be continuously recorded if desired. For obvious reasons, weighing lysimeters are more expensive than the nonweighing kind. It is also obvious that while the tanks may produce reasonably good values for water consumption by grasses and other small plants, useful values for larger plants

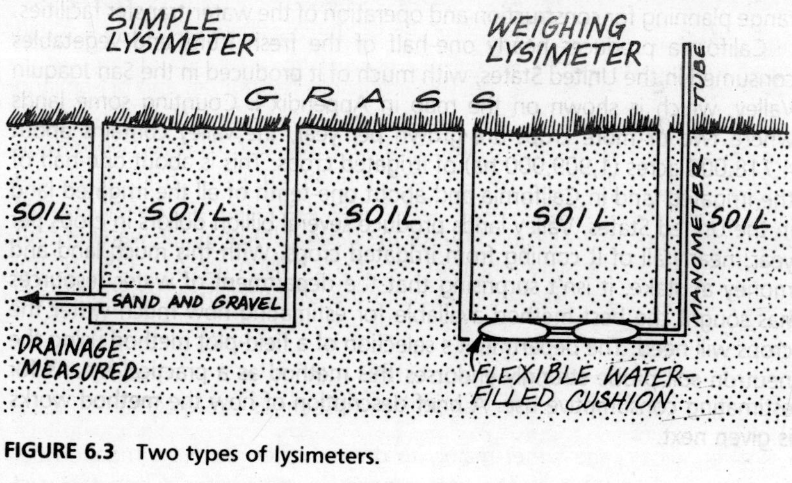

FIGURE 6.3 Two types of lysimeters.

such as mature trees will require some other method. The lysimeter method has other disadvantages as well. Probably most serious is the very small area of growing plants it measures. One would have more confidence in measurements made in larger areas—for example, real farmers' fields in which commercial crops are being grown.

Development of the neutron soil probe has made practical measuring periodic changes in soil-water storage beneath a growing crop, and thereby estimating evapotranspiration with enough precision to plan irrigation schedules in keeping with plant needs. Consequently, in recent years the emphasis in estimating evapotranspiration, especially in irrigation agriculture, has largely shifted from using lysimeters to using the neutron probe in field plots where natural conditions prevail. Chapter 5 explained operation of the neutron probe.

Neutron-Probe Method

A pioneer of the neutron-probe method, the California Department of Water Resources probably has done more intensive and longer field studies of evapotranspiration, or consumptive use, for agricultural crops than any other organization in the world. This effort is no doubt due to the need for those kinds of data in planning and building the California State Water Project—a truly monumental undertaking. With most of the natural precipitation and runoff in the northern part of the state and much of the need for water in the south, the state's water project has, through interbasin transfers, rearranged a substantial part of the natural water flow. In building large canals hundreds of miles long, the state needed to know how much water would be required and where. For more than two decades the department of water resources has performed field studies of consumptive use to aid in long-range planning for construction and operation of the water-transfer facilities.

California produces nearly one-half of the fresh fruits and vegetables consumed in the United States, with much of it produced in the San Joaquin Valley, which is shown on the map in Appendix I. Counting some lands where two or more crops are grown per year, the valley contains about 4,718,000 acres (1,909,000 ha) of irrigated crops. This is about one-half of the irrigated land in California and about one-tenth of all the irrigated land in the United States. Valley lands produce several billion dollars in crops per year, nearly all of it coming from irrigated crops. With this much land and money at stake, it isn't surprising that the department of water resources has sought the best method available for estimating how much water the crops will need. The department's selection of a field-plot method using the neutron soil probe strongly endorses this method as a practical means for estimating consumptive use. A brief description of how the method works is given next.

ESTIMATION OF EVAPOTRANSPIRATION FOR A GROWING CROP

The neutron-probe method studies field plots under actual working conditions, making measurements in commercially farmed fields and orchards. These studies, done with the cooperation of individual farmers, often continue for several years in a single plot. An investigation therefore includes several growing seasons for annual crops or several years of production for orchards or vineyards. In early studies, scientists had to take actual soil samples at various depths with soil tubes to monitor changes in soil-water storage, but in recent times they have made soil-water measurements with neutron probes, run in permanent, aluminum-cased access tubes.

Once a field site is selected and the farmer agrees to cooperate, land-and-water-use analysts, as these scientists are called, move in with drilling equipment and bore a number of holes cased with thin-wall, aluminum tubing 2 inches (5 cm) in diameter. The holes extend well below the root zone of the crop being studied. The casing has an expendable (and replaceable) upper portion about 18 inches (46 cm) long that may be destroyed when the farmer plows or otherwise cultivates the field. The holes are carefully located by surveys, and rubber stoppers are used to close both the surface opening and the part below 18 inches (46 cm), so researchers can recover the access tube at the beginning of each growing season. Figure 6.4 shows how an actual pattern of neutron-probe tubes was installed in a vineyard in the southern San Joaquin Valley. Their purpose was to determine the consumptive use of water by grape vines during the growing season.

Data gathered in the field provide an estimate of evapotranspiration through the following water-balance equation:

$$\text{Evapotranspiration} = \text{Precipitation} + \text{Irrigation} - \text{Runoff} \pm \text{Change in Soil-Water Storage} - \text{Deep Percolation}$$

"Deep percolation" refers to soil water that flows down beyond the root zone by gravity drainage.

Irrigation water applied to the field is carefully measured, as is any runoff (called return water or tail water) after irrigation. Precipitation is measured at a temporary weather station, called an *agroclimatic* station. These temporary weather stations are established near the field plots and are maintained for the period of investigation, usually three or four growing seasons. The agroclimatic stations include a Class A pan to measure evaporation during the term of the study, and sometimes instruments to measure other climatic data such as solar radiation, maximum and minimum temperatures, and so on (see Figure 6.5).

Changes in soil-water storage are measured periodically during the season by lowering a neutron probe into the access tubes. Figure 6.6 (page 128)

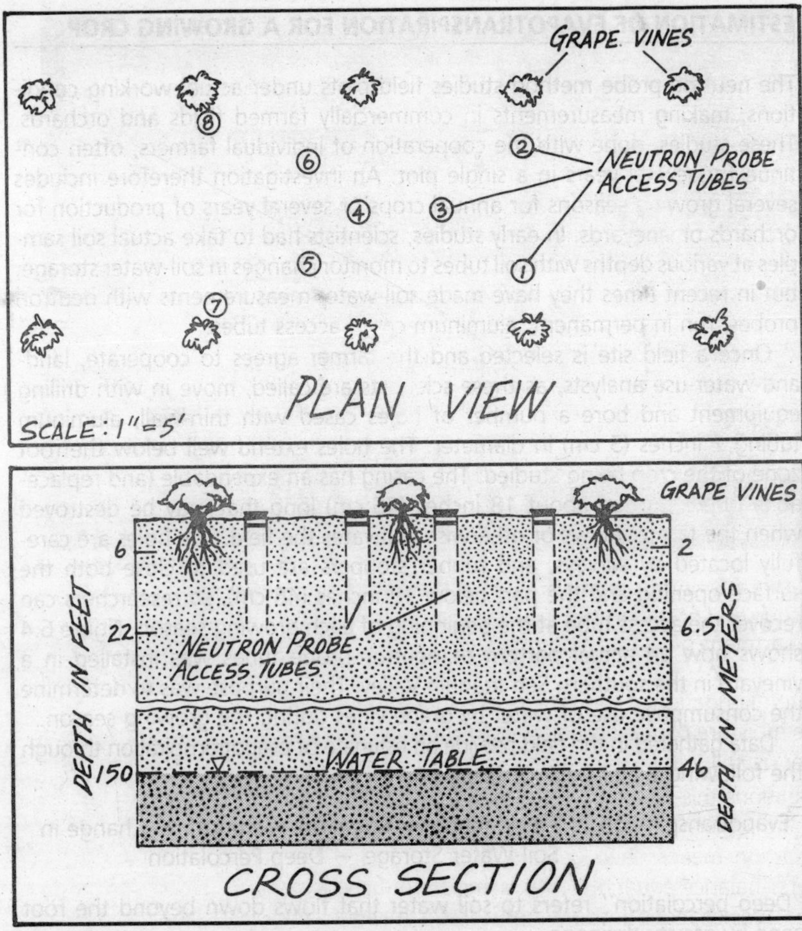

FIGURE 6.4 **Typical installation of neutron- probe access tubes in a vineyard in the southern San Joaquin Valley, California.**

shows a technician making neutron-probe readings from a platform elevated above the ground. The elaborate platform being carried by the technician in Figure 6.7 (page 129) prevents the observer from coming too close to the access tube while making measurements. Compaction of the surface soil could change infiltration rates, and that (or mechanical damage to the vegetation) could affect evapotranspiration rates from the crop.

At the beginning of the growing season irrigation water is applied to the field to bring the soil up to field capacity throughout the rooting depth.

FIGURE 6.5 Agroclimatic station of the California Department of Water Resources. Located in the southern San Joaquin Valley (photo: California Department of Water Resources).

Careful irrigation at this time results in a soil layer below the root zone where the moisture content is below field capacity. This so-called "dry" zone, however, is still within the depth measured by the neutron probe. During later irrigations this moisture-deficient zone can retain any deep percolation that would otherwise pass downward out of the measured soil profile. That retention means deep percolation is zero (and the term drops out of the water-balance equation). This is important because while the neutron probe measures changes in soil-water storage, it cannot measure water flow. With some initial experimentation and cooperation between the farmer and department of water resources scientists, usually the irrigation supply can be adjusted to permit only minimal deep percolation. If it were not minimal, and if it were not retained in the dry zone, deep percolation would be a major source of error in determining evapotranspiration by the neutron probe method.

In the vineyard shown in Figure 6.4, the grape vines were found to need 2.3 acre-feet of water per acre (7,010 m³/ha) for the average (April 15–October 30) growing season. The San Joaquin Valley contains more than 500,000 acres (202,350 ha) of vineyards, and even though consumptive use may vary in different locations, these figures indicate that grape crops use more than

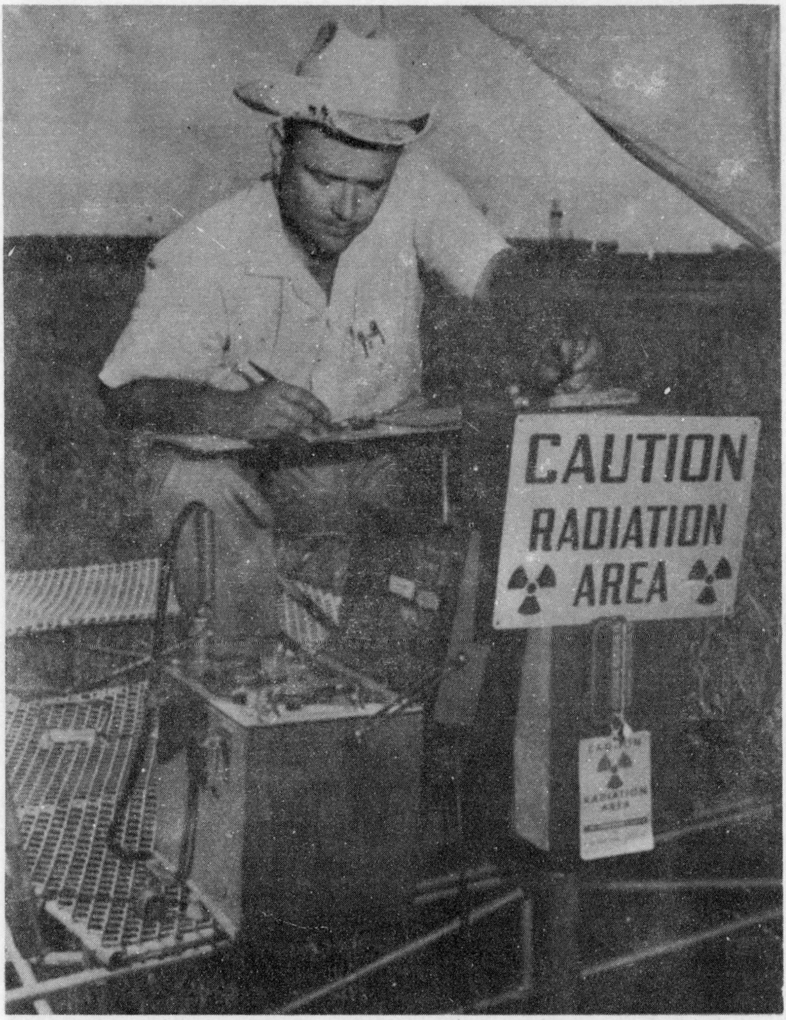

FIGURE 6.6 DWR technician reading and recording neutron counts from the probe in the subsurface. The brass cable stops seen in the technician's left hand are spaced along the cable so that the probe can be located opposite the same soil regions each time the probe is lowered into the access tube. The large box through which the cable runs is a shield for the probe when it is not in use: it prevents exposure to radiation from the probe's radioactive source. The instrument case to the left of the shield contains a scaler counter that records the number of slow thermal neutrons being received by the detector in the probe. Before going into the field the instrument is calibrated so that the readings can be converted to percentages of soil moisture (photo: California Department of Water Resources).

FIGURE 6.7 Sampling platform used with the neutron probe. Working from this platform, the observer never walks on the soil surface within 6 feet (2 m) of the neutron-probe access tube. This avoids soil compaction and damage to the vegetation at the sampling site. Platform is shown being relocated from one tube to the next (photo: California Department of Water Resources).

1 million acre-feet (1,233,500,000 m³) of water each year. (Do you think an important relationship might exist between the water cycle and a "wine cycle"?)

Field-plot studies are expensive and time consuming. Only a few have been done, generally one for each of the major crops grown in the San Joaquin Valley. By correlating the estimated consumptive use with Class A pan records from the nearby agroclimatic stations, though, coefficients have been determined empirically for each crop. These indicate water needs throughout the growing season and allow estimation of consumptive use for that particular crop at other locations. Long-term investigations have demonstrated the practicality of such crop coefficients. With a Class A pan record in a given locality, we can estimate consumptive use of water at that place for most of the major crops grown in California. Simply multiplying the Class A pan data by the crop coefficient produces the expected evapotranspiration, just as the pan coefficient determined by the Lake Hefner study (outlined in Chapter 3) produces the evaporation estimates for lakes and reservoirs.

If the results of the research on crop water use were applied statewide, an estimated real, on-farm savings of as much as 600,000 acre-feet (740,100,000 m^3) could result per year. That wouldn't affect overall operation of the water cycle, but it would release water now going to irrigation for other uses, resulting in a substantial saving of energy. In the long run, more efficient use of water might even lower prices for food and fiber grown on irrigated farms.

POTENTIAL EVAPOTRANSPIRATION

This chapter has emphasized how reliable estimates of evapotranspiration require considerable time, effort, and expense. Therefore, devising a formula or equation using readily available data to estimate evapotranspiration would greatly simplify the work of the hydrologist, soil scientist, or irrigation engineer. This approach seems to have a strong appeal to those people who are inclined toward neat, mathematical solutions for otherwise difficult natural problems. Such a philosophy resulted in the concept of potential evapotranspiration.

Potential evapotranspiration is the total water loss occurring from a short green crop (grass), extensive in area, that completely shades the ground and is never short of water. By definition it cannot exceed evaporation from a free water surface subjected to the same climatic conditions. You could measure potential evapotranspiration at a specific location by planting a lysimeter with grass and keeping the soil in the tank moist at all times. To be effective, the lysimeter would have to be located amid an extensive area also planted with grass. A lysimeter used in this way is sometimes called an *evapotranspirometer.*

The more common way of obtaining potential evapotranspiration is using one of several equations that have been proposed for this purpose. These equations contain various meterological data such as solar radiation, humidity, temperature, and wind. Latitude is also an important factor, as that will determine the duration of sunshine during the day and hence the amount of radiation expected from a clear sky. Cloud cover may also affect the radiation supply. These are the same environmental conditions that control evaporation from a Class A pan.

Since nature seldom follows precise mathematical formulations, it is not surprising that real evapotranspiration nearly always differs from the computed potential evapotranspiration. The scientists who first proposed the idea were working in the humid climates of the northeastern United States and England, where computed potential often approximates real evapotranspiration. In the arid or semiarid climates of the Great Plains, or western states, however, computed and measured values may differ substantially. One reason for this is potential evapotranspiration's inability to exceed open water evaporation. Experiments in the Great Plains have shown that measured evapotranspiration is sometimes larger and sometimes smaller than evaporation from a nearby open water surface (Class A pan). In arid and semiarid regions, therefore, predictive equations have often been empirically modified to account for the hot, dry conditions.

If potential evapotranspiration often is such an uncertain quantity, why mention it all? A number of books, including some listed at the end of this one, address potential evapotranspiration, providing a whole array of equations for computing it. For the mathematically inclined, especially those who find pleasure in working with computers, these equations may have a certain appeal. But if you want a quick and easy approximate value for evapotranspiration, the data from a Class A pan, with the appropriate pan coefficient, will usually suffice. Where the pan data can be correlated with real, field-plot measurements (as in the California department of water resources method), the estimate of evapotranspiration probably will be as close as possible using current estimation methods. The point is that whenever you have the meterological data needed to apply to one of the equations, you will very likely have Class A pan data, too. Here, as in most areas of natural science, the simpler the approach (provided it is physically sound), the more reliable the estimate.

EFFECT OF EVAPOTRANSPIRATION ON WATER SUPPLY

Changes in Runoff

When transpiring vegetation pumps large quantities of soil water or groundwater into the atmosphere, is that good or bad? It depends, of course, on the purpose of the vegetation. Evapotranspiration always puts water back into the atmosphere where we can no longer use it, but if that moisture loss produces a crop of food or fiber, we might consider that a fair trade. On the other hand, when the "lost" water comes from a potential supply that we wanted to use, we might be disturbed and try to do something about it.

In regions where water supply nearly balances demand, any water lost to the system may be a serious matter. Some urban areas depend on distant watersheds to supply all their water, and in recent years population growth has increased demand for water. Engineers who run water systems have tried several methods to increase supply such as building dams and reservoirs, extending supply lines farther into the hinterland, and even cutting out many of the trees and reducing brush cover in the watersheds.

It used to be thought that the more trees the better, but long-term runoff measurements on drainage basins from which many of the trees (and/or brush) have been removed have shown a marked increase in water yield after the vegetation was removed. A study several years ago in a Southwest mountain watershed found that removing only about 15% of the forest trees caused a sizable increase in stream flow over a period of years. The increased runoff occurred without a significant increase in sediment load in the streams, indicating that tree removal caused essentially no increase in erosion.

In addition to the number of trees in a watershed, the *kind* of trees also may influence water yield. Transpiration rates differ widely among the different species of trees. In a long-term experiment in two Southeast watersheds, experimenters converted a deciduous hardwood forest to white pine. They found that simply changing tree species reduced stream flow almost 20% annually by the end of the 15-year study. Lower runoff occurred largely because of the higher annual transpiration from the pine trees and also because the pine trees intercepted more precipitation during the year than the hardwood, which lost its leaves in the autumn.

Changes in Groundwater Storage

Transpiration, in addition to affecting surface runoff, may also directly affect groundwater storage. When the water table is within reach of plant roots, all or part of the water transpired to the atmosphere may come directly from the groundwater reservoir. In arid regions this sometimes results in substantial water losses. In the mid 1920s the United States Geological Survey ran a series of experiments in Utah's Escalante Valley, an arid region with a shallow water table. Shallow wells were installed in fields of natural vegetation, in alfalfa fields, and in bare fields from which the vegetation had been cleared. Figures 6.8 and 6.9 show fluctuations in the shallow water table due to transpiration by growing vegetation, both in native brush and in cultivated alfalfa. Also shown is a water-table recording from beneath the bare ground where no transpiration was taking place.

Data for the graphs of Figures 6.8 and 6.9 were taken from automatic water-stage recorders placed on shallow wells in the designated fields. The graphs show the daily fluctuation of the water table due to uptake of groundwater by the plants. The downward trend of water levels in Figure 6.8 also shows the short-term, net depletion of groundwater storage in the general region of the transpiring vegetation. Figure 6.9 shows how plant growth and plant bulk markedly affect transpiration rates. When the alfalfa hay was

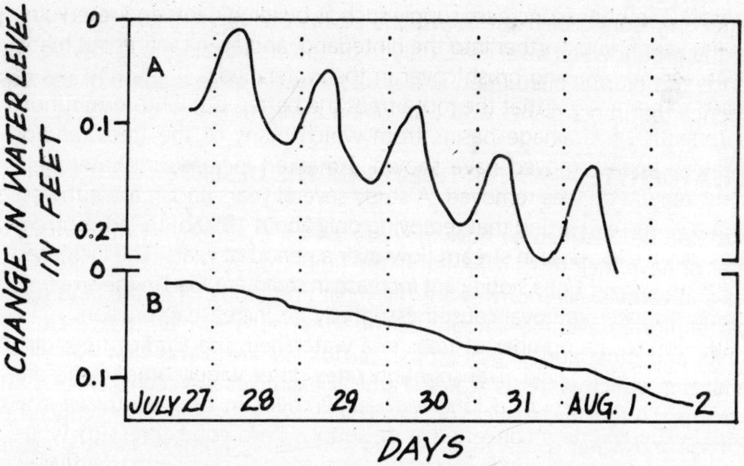

FIGURE 6.8 Comparison of daily fluctuations of the water table—(a) beneath fields of native brush; and (b) beneath cleared land—for the period July 27 to August 2, 1926. Located in Escalante Valley near Milford, Utah (source: U.S. Geological Survey Water Supply Paper 659-A, 1932, 43.).

cut, less plant surface was there to transpire, and the water table under the field temporarily rose.

Another interesting feature of these graphs is the plants' daily fluctuation in groundwater uptake. Since photosynthesis and the associated transpiration

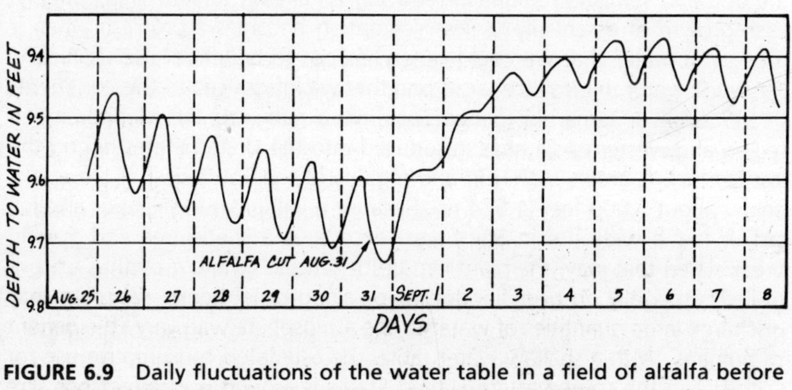

FIGURE 6.9 Daily fluctuations of the water table in a field of alfalfa before and after alfalfa was cut, August 26 to September 8, 1926. Located in Escalante Valley near Milford, Utah (source: U.S. Geological Survey Water Supply Paper 659-A, 1932, 45.).

occurs only during daylight hours, the plants use very little water at night, and the groundwater level rises, only to be lowered again the following day. In fact, this diurnal fluctuation in a shallow water table is some of the best evidence that most plants stop transpiring at night.

PHREATOPHYTES—WATER THIEVES OF THE DESERT

Plants can be classified by how they use water. *Hydrophytes* either grow in water or have their roots in water all the time. Common examples are water lilies and cattails. *Mesophytes* are the most numerous. They are land plants that require an average amount of moisture, such as pine, oak, hemlock, maple, and most agricultural crops. *Xerophytes* are plants that live in arid climates in which water is scarce. Familiar examples are cactus, yucca, and sagebrush. A fourth classification, used by hydrologists and others concerned with soil and water conservation in arid and semiarid regions is the *phreatophytes.* The word has Greek roots meaning "well plant," an appropriate description of the way these plants function. They send roots down to the water table, literally "pumping" water from the groundwater reservoir into the atmosphere. Figure 6.10 illustrates the distinction between xerophytes and phreatophytes.

The classification has not been generally adopted by botanists because most of the so-called phreatophytes can also be grown where rainfall or irrigation water is sufficient to allow plant growth without recourse to the water table. In humid regions, many plants—especially trees—use both soil water and groundwater. Phreatophytes may become a problem mainly in the arid and semiarid regions in reducing an already limited water supply. Lush growth of essentially useless vegetation becomes a problem when it robs us of water that we could otherwise put to beneficial use. Ironically, the worst water thief, saltcedar, is one that we introduced to this continent.

Saltcedar, or tamarisk, (genus *Tamarix*), a native of the Mediterranean region and western Asia, was introduced into the United States during the last century. It grows mainly in arid regions of the Southwest at elevations below about 5,000 feet (1,524 m). Because it will grow in saline or alkaline soil, it found wide use in Southwest deserts as a shade tree and a wind break. Used that way, the plant can be beneficial. Where growth is uncontrolled, especially on river floodplains or along canal banks, saltcedar may discharge large quantities of water to the atmosphere with very little benefit to humans. With a shallow water table, the estimated consumptive use for saltcedar in the southwestern United States is around 6 acre-feet per acre per year (18,288 m^3/ha/year). Where the water table is deeper (more than 7 feet or 2 meters), annual water use is somewhat less. Where saltcedar can extend its roots to the water table, however, it can transpire prodigious quantities of groundwater into the atmosphere. The roots can go down a

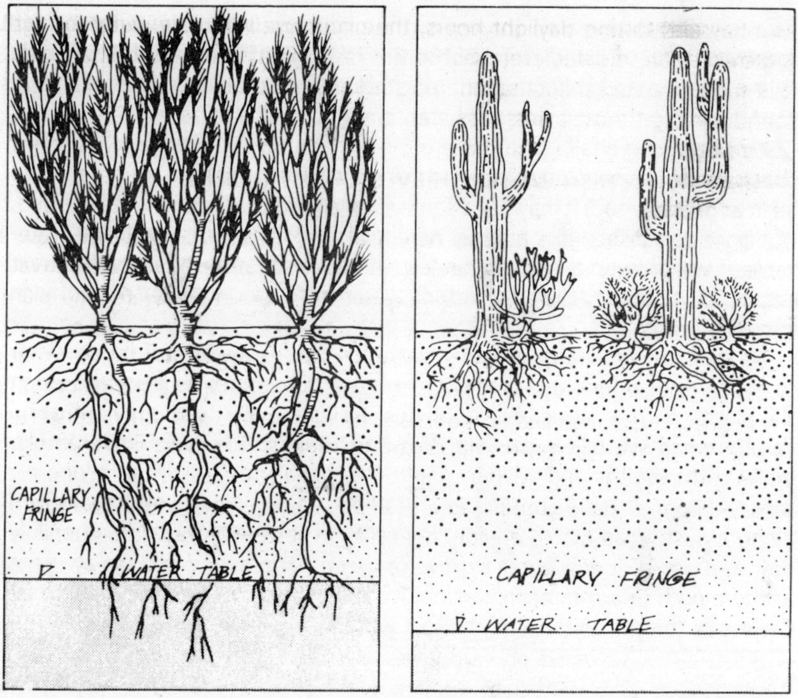

FIGURE 6.10 Comparative relationship of phreatophytes and xerophytes to the underlying water table (source: U.S. Geological Survey).

long way, too. An excavation of the Suez Canal found saltcedar roots at a depth of 30 meters (almost 100 feet).

Common Phreatophytes

While saltcedar is sometimes an undesirable alien in the U.S. natural plant community, some of our native phreatophytes are also heavy water users. Willow (*Salix*), greasewood (*Sarcobatus*), saltgrass (*Distichlis*), rabbitbrush (*Chrysothamnus*), and pickleweed (*Allenrolfea occidentalis*) are common in the Southwest, and all are associated with high water consumption. Some of the larger trees, such as alder (*Alnus*) and cottonwood (*Populus*) are also classed as phreatophytes. Plant communities like these are partly responsible for the large evapotranspiration and small runoff in arid regions

The most common cultivated phreatophyte is alfalfa (*Medicago sativa*), also called lucerne. Introduced in the United States from Europe in the middle of the eighteenth century, by the mid-nineteenth century it was established as a hay and forage crop throughout the country. In the arid West, alfalfa is grown as an irrigated crop; but in the East irrigation is seldom practiced. It is a deep-rooted plant—in one documented case in Nevada, roots were found coming through fractured rock in the roof of a mine tunnel 129 feet (39 m) below an alfalfa field. Where the water table is within reach of plant roots, alfalfa has been established in fields cleared of worthless phreatophytes such as greasewood. It may need some supplemental irrigation to get started, but once established it is a hardy perennial, and if its roots reach the water table, it will grow in a field indefinitely. In such a situation, alfalfa will extract about 36–42 inches (90–107 cm) of water from groundwater during each growing season.

At least in local areas, evapotranspiration causes the loss of much water to further use. In low-rainfall regions, this may sometimes be regretted when the plants, such as some phreatophytes, are essentially useless. We must be careful when we use words like *loss* and *useless,* however. For example, while cottonwoods and willows may be essentially useless in an economic sense, they may have considerable aesthetic value. When you have driven along a dusty road across a wide, dry valley in eastern Nevada on a hot July day, and you suddenly enter a grove of green cottonwood and willow trees at the mouth of a mountain canyon, it's difficult to consider the trees to be useless or the transpiration to be lost water.

APPLICATIONS

An obvious application for evapotranspiration is, of course, irrigation agriculture. We have been applying water to land through irrigation for thousands of years. Good-quality water, usually river water, is transported to land for the purpose of growing useful food or fiber crops or for irrigating pastures for grazing animals.

Are these the only applications for evapotranspiration? How about irrigating vegetation with poor-quality water, thereby using evapotranspiration to dispose of unwanted wastewaters? Of course, you would need to select plants that could tolerate whatever is in the wastewater, and you may or may not get a useful crop. If the process proves to be the least costly method of waste disposal for the given water, then it may be worthwhile even if it only produces a lush growth of weeds.

Evapotranspiration as a Method for Waste Disposal

Throughout most of human history water supply problems have involved getting the water and conveying it to the point of use. Except in the case of flooding, disposing of unwanted water was never much of a problem until recent times. Now, however, disposal of wastewater from sewage plants and industrial processes creates problems that our ancestors didn't have to face.

We are all aware of "beneficial use" of natural resources and the widely accepted principles of protecting and enhancing the environment. Did you know that until just a few years ago another widely accepted idea was that the most beneficial use for rivers and streams was as natural sewers to carry away the wastes of our so-called advanced civilization? Considering the recent growth in environmental consciousness, that statement may seem incredible. Nevertheless, it is true, and if you doubt it, do a little library research and find out how many city sewage works and industrial plants were discharging raw sewage and untreated wastewaters into the nation's rivers just 20 years ago.

In some cities at that time, a glass of water drawn from a tap might occasionally include a "head" of foam. This was foaming detergent from a sewage outfall, located upriver somewhere from the water intake for the city in question. Even though it was disinfected with chlorine and contained no pathogenic organisms, it wasn't easy to drink that water, knowing what its history had been. If you visited these areas today, you would find much better water coming from the tap. The cities and towns upriver now discharge treated effluent into the river.

Land Disposal of Sewage Effluent. Sewers have existed for a long time—in fact, almost 6,000 years. Archaeologists studying the remains of a sewage system at Nippur, India, thought it was probably built in about 3700 B.C. Near Baghdad, in present-day Iraq, there are remains of sewers constructed in the 26th century B.C. Rome had an extensive, efficient system of sewers, as did other ancient cities.

In the past, sewers were simply means for collecting and conveying domestic wastes out of the city to some point of disposal beyond the view (and smell!) of the inhabitants. No thought was given to treating the wastes. They were just flushed away downstream into the nearest river. Only within the last century have modern cities provided even the most rudimentary treatment for their sewage effluents. In the beginning, and until fairly recent times, treatment consisted mainly of removing only the solids in sewage and discharging the remaining effluent to rivers or to the land. Sewage effluent was diluted by the river water, and on the way downstream natural biological processes gradually eliminated wastes in the stream. Similar biological processes gradually purified effluent that was put out on the land in "sewage farms."

Sewage farming, which began in the United States around 1880, is still practiced in a number of cities, mostly in the arid Southwest. The practice serves a dual purpose: The effluent is purified by natural biological processes in the soil, and some of the water is used in growing commercial crops.

Water not used in evapotranspiration passes downward by deep percolation to the underlying groundwater body. Used this way, sewage farming is the final step in treating sewage effluent and reclaiming the water for other uses.

Most states restrict crops grown by sewage irrigation to those not consumed directly by the public, such as cotton, pasture of beef cattle (not milk cows), alfalfa, cereal grains, and other nonedible crops. If the effluent has undergone advanced treatment, as it has in many present-day treatment plants, the water may be used on a wider range of crops, but still not on crops that would be consumed uncooked. This method for sewage disposal and water reclaimation is used most in arid or semiarid regions in which water is chronically in short supply. In the southern San Joaquin Valley of California, for example, about 75% of municipal and industrial wastewaters is reclaimed for use, much of it by irrigating crop lands in municipal sewage farms.

Even when sewage farms are effective in cleaning up and reclaiming water, the fraction of effluent that passes down to groundwater still has one undesirable feature. Most organic pollutants are eliminated by biological processes in the soil, but much of the mineral material dissolved in the water is unaffected by these processes. Dissolved constituents from domestic sewage are not toxic; nevertheless, in time mineral content of the underlying groundwater can build up. This is not apt to become a serious concern for many years, but eventually it will degrade the groundwater resource. This is just one more example of the price we must pay for what we do to natural systems.

Waste streams leaving an industrial process, however, may contain dissolved constituents that are not as innocuous as those in domestic sewage. Even though the water will support certain kinds of vegetation, some constituents in solution may be toxic to humans or to commercial crops, rendering the water undesirable as a source for groundwater recharge. In this situation the disposal facility must pass essentially all the wastewater into the atmosphere through evapotranspiration and none downward through deep percolation to the groundwater body. This is possible, as the water-disposal facility described next demonstrates.

Disposal of Oil-Field Wastewater. Most oil wells produce water along with oil. After separation at the surface, the oil goes to a refinery and the water goes to some sort of waste-disposal facility. The chemical quality of these oil-field waters ranges from very salty to almost fresh.

In the southern San Joaquin Valley of California many wells produce oil in the midst of highly productive irrigated farmlands. Most oil-field waters from these wells are far too salty for irrigation and must be disposed of in other ways. Subsurface injection into deep geologic formations or disposal on the surface through evaporation or evapotranspiration are the chief methods used.

The evaporation ponds discussed at the end of Chapter 3 are typical of those used for water disposal by straight evaporation. Ponds like these are usually necessary for oil-field waters with a heavy load of dissolved salts, which would be unsuitable for most crops. For more dilute waters suitable for certain kinds of vegetation, wastewater may be disposed of by evapotranspiration rather than by straight evaporation. Where it can be used, evapotranspiration may actually enhance the arid environment of the disposal site and is therefore aesthetically more desirable.

One such disposal facility serves the Race Track Hill area of the Edison oil field, located in the southern San Joaquin Valley about 7 miles (11 km) east of Bakersfield, California (see map in Appendix I.)[1] Much of the oil-field surface area is planted with high-value crops, chiefly citrus and grapes. Historically all irrigation water was pumped from wells. Now, surface water from the Arvin-Edison Water Storage District serves some of the area. In dry years, however, all the farmers may still have to depend on groundwater for an irrigation supply.

The oil-field water is only moderately saline, about 12% as salty as sea water, but it contains far too much of one constituent that is toxic to the area crops. The troublemaker is boron, which is essential in trace amounts to most plants but extremely toxic to some plants in greater amounts. Irrigation water with as little as 1–2 milligrams per liter of boron can reduce yields in citrus and grapes. Water with several milligrams per liter of boron may even be fatal to the trees or vines themselves. The Race Track Hill oil-field water contains around 17–20 milligrams per liter of boron and thus must not be allowed to mix with water used to irrigate the vineyards and citrus groves. This is not easy, since trees or vines surround many of the wells. To ensure safe disposal, the well operators have constructed a completely enclosed pipeline system to bring the water from the oil-water separators at the wells to a central cleaning plant that removes the last vestiges of oil. The water is then pumped to the final disposal site, located about 4 miles (6.4 km) away in the Sierra foothills.

The disposal site consists of 240 acres (97 ha) of barren, hilly land with an average elevation of about 1000 feet (305 m) above sea level and with a topographic relief of about 300 feet (91 m) within the site area. Annual rainfall at the site averages a little over 5 inches (127 mm) and evaporation (measured with a Class A pan) is around 80 inches (2032 mm) per year.

---

[1]The operator of the Race Track Hill disposal site is the Valley Waste Disposal Company (VWDC), a nonprofit company controlled by oil companies producing from fields throughout the southern San Joaquin Valley. The author of this book was a consultant to VWDC for many years and appreciates the company's permission to publish these facts about the disposal operations at Race Track Hill.

Although the desert soils show poor profile development and are relatively impervious, trees and grasses grow readily wherever enough water is available.

Figure 6.11 is an aerial photograph of the general layout of the site. In addition to the irrigation-type sprinkler system that delivers water to the plants, the area includes a number of shallow sumps, or basins. These were constructed by bulldozing small check dams in the natural drainage courses. The sumps have a combined storage capacity of about one million barrels (42 million gal or 159,000 m³) and are used to store wastewater during winter months when evapotranspiration is at a minimum.

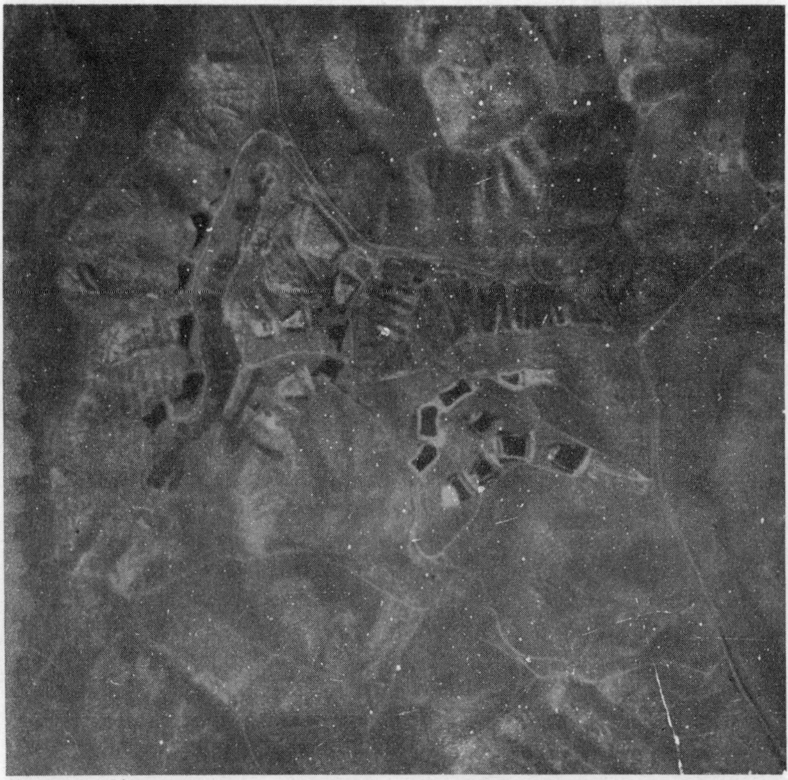

FIGURE 6.11 Aerial photograph showing evaportranspiration disposal site for Race Track Hill waste water. Sprinkler lines are clearly outlined by darker areas of growing vegetation. The dark sumps contain water. Dry sumps appear lighter than surrounding ground. Borders of saltcedar trees can be seen around several of the sumps.

Water is pumped from the cleaning plant in the oil field to the receiving sump at the top of the hill, from which gravity flow feeds it to the sprinkler lines or to the storage sumps. Some 800 sprinkler heads (see Figure 6.12) spaced 30 feet (9 m) apart each discharge about 2 gallons per minute (0.13 L/sec) from $\frac{1}{8}$-inch (3.2 mm) orifices at a pressure of about 36 pounds per square inch (248 kPa). The site was designed originally for disposal of 20,000 barrels (840,000 gal. or 3,180 m^3) per day, and in its first year of operation, 1960, the daily average ran a little over 18,000 barrels (756,000 gal. or 2860 m^3) per day. Over the years, wastewater volume has declined as production from the oil field has declined, and by 1984 the disposal volumes were running around 14,000 barrels (588,000 gal. or 2225 m^3) per day.

Operation of the system is predicated on obtaining about a 6–8-inch (15–20 cm) penetration of water in the soil, then moving the flow to another sprinkler line. Since the sprinkler lines themselves are not moved, operation of the system is very simple. Operators who also oversee the water-cleaning plant in the oil field simply change the flow from one sprinkler line to another as required by plant and soil-moisture conditions.

As you can see from Figure 6.11, these dry desert hills have little native vegetation. It was obvious from the beginning that native vegetation on the site could not do the job alone. Saltbush (*Atriplex polycarpa*) and some native rye grasses have survived, but most of the native grasses and shrubs were not able to tolerate the saline, boron-rich wastewater. In an effort to find plants suitable for these harsh conditions, test plots were established. Plant species that survived and prospered were selected for planting around the

A. B.

FIGURE 6.12 (a) Well-established stand of tall wheatgrass along a sprinkler line. (b) Close-up view of sprinkler head in tall wheatgrass.

sumps and along the sprinkler lines. These include bermudagrass (*Cynodon*), saltcedar or tamarisk (*Tamarix aphylla*), tall wheatgrass (*Agropyron elongatum*), birdsfoot trefoil (*Lotus corniculatus*), alkali wild ryegrass (*Elymus triticoides*), and alkali sacaton (*Sporobulus airoides*).

The plant species most successful at coping with the high boron concentration have been tall wheatgrass, saltcedar, and bermudagrass, the species that represent much of the lush vegetation seen along the sprinkler lines and around the sumps in Figure 6.11. Figure 6.12 shows a field of tall wheatgrass along a sprinkler line. A wastewater sump with a border of saltcedar trees is shown in Figure 6.13. A general view of part of the disposal site appears in Figure 6.14, which also shows the contrast between unirrigated desert land (in the foreground) with the luxuriant growth of trees and grass being irrigated with wastewater.

Where plants are growing under irrigation, the roots and associated biological activity keep the surface soil open and pervious. Analyses of shallow soil samples show that soil in the root zone of the grasses seems to have a stabilized content of about 18–32 milligrams per liter of boron. Some salt

FIGURE 6.13 Saltcedar trees around a wastewater sump.

FIGURE 6.14 Portion of disposal site showing contrast between unirrigated land in foreground and irrigated vegetation in background. Trees are salt-cedars and most of the grass is tall wheatgrass. The white wispy lines above the grass are water sprays from the sprinklers.

may be accumulating below the root zone. Under the present operation, however, it appears that salts are effectively buried beneath this shallow zone. As long as water is applied at a rate more or less balanced by evaporation from bare soil and transpiration from the plants, very little seepage water will be available for deep percolation.

No evidence shows that significant quantities of water are percolating downward from the storage sumps. During construction, heavy equipment scraped off the surface soil and compacted the subsoil. Sump bottoms are now relatively impervious, and water disappears from the sumps at a rate more consistent with expected evaporation losses rather than with percolation. The main problem to avoid is surface runoff out of the disposal site, and so far this has not happened.

For almost a quarter of a century, oil-field water has been disposed of safely and efficiently by evapotranspiration at the Race Track Hill disposal site. Unlike many disposal sites, where wastewater has degraded the natural environment, operations at this site have substantially upgraded the environment. This is plain from the photos that show vegetation growing under wastewater irrigation as contrasted with surrounding areas of unirrigated, unaltered desert land.

Several years after operations began, a series of experiments were conducted to show the suitability of the disposal site environment for animals as well as for plants. It was found that cattle, horses, and sheep would graze the grasses irrigated by wastewater. The cattle gained weight less rapidly than cattle grazed at other nearby ranges, but the sheep seemed to thrive on this feed as well as did sheep on the neighboring ranges. No deleterious effects have resulted from drinking the water or eating the grasses.

One of the storage sumps that holds water year-round was selected for a fish experiment. Young bass (*Roccus saxatilis*), crappie (*Pomoxis annularis*), and bluegill (*Lepomis macrochirus*) were planted in this sump. They grew rapidly and provided a sport for the disposal-site operators, who harvested most of them with rod and reel. Migratory water fowl that frequent the ponds in season also ate some of the fish. A permanent population of mosquito fish (*Gambusia affinis*) has dwelled in this same pond for several years.

Improved conditions for birdlife are also evident at the site. In addition to the migratory water fowl, several hundred quail have found food and cover in the vegetation grown with wastewater.

SUMMARY

Where climatic and soil conditions are favorable, evapotranspiration may be an efficient method for wastewater disposal; however, evapotranspiration is not a viable alternative for disposal of any wastewater at any location. Conditions at the Race Track Hill site are especially favorable. Not only are these desert soils relatively impervious, especially a few feet below the root zone, but little usable groundwater exists beneath the site. The groundwater basin serving the farmers in the valley is isolated from the site by geological faults that effectively separate groundwater in the two areas. Even in the absence of these fault barriers, operation of the site would preclude all but a minimal downward percolation of waste water. Without the faults, though, installation of monitoring wells might be needed to ensure that disposal operations were not polluting usable groundwater. Monitoring wells were not necessary but could be installed without difficulty where needed.

7 Groundwater

And Isaac's servants digged in the valley, and found there a well of springing water.

(Genesis 26:19)

Although we have known about groundwater for thousands of years, it has nevertheless remained a subject of mystery and speculation. Early humans probably learned from the animals how to find water beneath a dry stream bed. A thirsty coyote in North America or a dingo in Australia will dig a drinking hole in a stream bed in which groundwater lies just beneath the surface. Primitive people learned to do the same thing, and after developing a pastoral economy, they learned to dig wells to supply their flocks (as reported in Genesis, above). A few of these wells are still in use after thousands of years. Even though much of the water supplies for most large cities come from surface streams and lakes, a large part of the world still depends on groundwater from wells and springs. In the United States, most rural and many suburban areas depend on wells.

No one knows precisely how much groundwater exists in the world; comparative estimates in Table 1.1 indicate that probably at least 200 times the volume of annual runoff from the world's rivers is stored as groundwater beneath the land surface. This enormous body of water plays a crucial role in the operation of the water cycle. In other parts of the cycle, water is stored for longer or shorter periods; a few days in the atmosphere, probably a few weeks or months in the soil, and a few days or weeks in rivers on the way back to the ocean. Once water enters the zone of saturation and becomes groundwater, however, it may stay there for a very long time.

Physicists have devised methods of measuring the age of groundwater issuing from wells or springs. The measurements are based on the rate of radioactive decay of certain molecules or ions held in solution, and the "age" is the number of years that have passed since the particular drop of water fell from the sky and entered the soil. During the past 25 years, age determinations of groundwater from all the continents have provided a new dimension for the groundwater reservoir—a dimension of time. Some water issuing from springs or wells has been underground only a year or two. Other samples have spent several decades in the ground. Some water from beneath the deserts of North Africa or Central Australia is so old that even the radioactive "clock" has run down. That groundwater probably infiltrated from rain that fell during the Pleistocene epoch, many thousands of years ago.

THE GROUNDWATER RESERVOIR

Groundwater reservoirs are the great regulating reservoirs for the land phase of the water cycle. They keep the rivers flowing long after the rain has stopped and the snow has melted. Although these reservoirs may store water to a depth of several thousand feet below the surface, the lower parts often represent essentially dead storage. Most of the natural flow happens near the top of the saturated zone, where groundwater is close to the surface. Under natural conditions the reservoirs are usually full to the spill point, and water moves continuously from areas of natural recharge to areas of natural discharge.

Areas of Natural Recharge and Discharge

Figure 7.1 shows the natural flow system. The diagram shows that in addition to recharge from infiltration of precipitation directly into the soil, streams flowing above the water table may also provide recharge. These are called *losing,* or *influent,* streams because they are losing water to the ground. Natural discharge may occur by evapotranspiration through phreatophytes (Chapter 6), by surface seeps and springs, and by seepage into streams. In humid regions most streams flow at or near the water table, and most natural discharge occurs through groundwater flowing into the banks or beds of the streams. Such streams are known as *gaining,* or *effluent,* streams because they are gaining water from the ground.

While the water volume depends primarily on climate, areas of natural recharge and discharge usually result from local geologic conditions. In humid regions more water infiltrates the soil and seeps down to groundwater than in arid regions, in which the opportunity for evapotranspiration is greater. In Australia, where evapotranspiration uses about 87% of the annual rainfall,

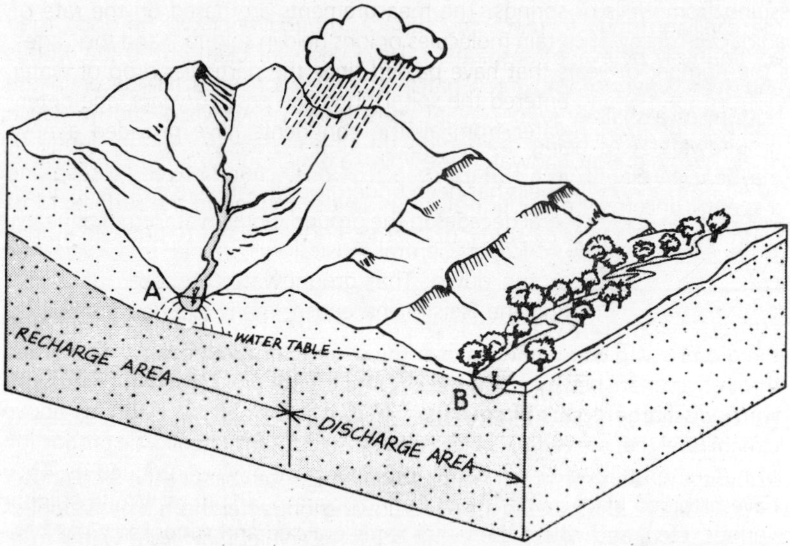

FIGURE 7.1 Groundwater flow under natural conditions.

on the average, only an estimated 1% of the rainfall ever reaches groundwater level. Conditions are better in North America. In the conterminous United States, the average annual increment to groundwater is about 3 inches (76 mm), or about 10% of the average annual precipitation. This amounts to about 400 billion gallons per day (1,500,000,000 m³ per day) of natural recharge, balanced by an equivalent amount of natural discharge, mostly to streams. Groundwater thus contributes about 35%–40% of the total water discharged by streams in the United States each year.

Recharge to groundwater will occur wherever enough rainfall, snowmelt, or surface streams infiltrate the soil to allow gravity drainage through the soil to the water table. The depth of the water table below the surface usually offers a good index of how effective recharge is at that particular locality. In humid regions the water table is generally only a few feet beneath the surface; in arid regions it is much deeper, and in some desert basins the water table is a thousand feet (300 m) or more below the surface. Even here, however, groundwater often moves from areas of recharge toward areas of discharge. Recent studies in the Basin and Range province of the western United States have demonstrated that some of the large desert springs are discharging water that entered the ground many miles away, sometimes on the other side of an intervening mountain range.

Groundwater discharge occurs wherever the water table nears or actually intersects the ground surface. If the water table is shallow enough to allow the capillary fringe to reach the surface, groundwater will be discharged continuously by evapotranspiration. Have you ever been out in the woods and seen a very moist, green patch of vegetation on a hillside or in the bottom of a shallow swale? What you probably saw was a spot in which groundwater was being discharged through a shallow capillary fringe or maybe even directly in a surface seep. The distinction between a seep and a spring concerns the amount of groundwater coming to the surface. Most hydrologists would say a *spring* discharges enough water to produce a small rivulet.

Springs. Springs occur in two ways. They may issue from an opening in the ground with water flowing over the surface, or groundwater discharged through a stream bed may flow directly into the stream. Most natural groundwater discharge occurs in streams, where it is invisible. The springs above stream level that are easily visible probably account for a negligible proportion of natural groundwater discharge, but they have always intrigued us: They have provided place names (Palm Springs, Saratoga Springs, White Sulphur Springs, etc.), and where the water is mineralized and supposedly has beneficial qualities, they have been called *spas,* after the famous mineral spring at Spa, Belgium. Because the geological conditions that produce springs may take many forms, springs have been classified on the basis of geology, type of opening, kind of water, and so on. A few common types of springs are shown in Figure 7.2. The existence of springs proves that the underground reservoir is full to the spill point, and water storage is at its maximum.

AQUIFERS

While the entire zone of saturation is referred to as the *groundwater reservoir,* it is seldom a single, homogeneous geologic formation. Usually a variety of rock types are present at any given location, and even though they may all be saturated, they often have widely varying hydrologic properties. Some would be called aquifers and others would not. The term *aquifer* comes from two Latin words—*aqua,* meaning water, and *ferre,* to bear.

To be called an aquifer, a geologic formation must be porous and permeable. It must store, transmit, and yield significant amounts of water to springs and wells. That definition is suitably vague, because no one has ever defined precisely how much water is "significant." But if you think about it, you can see that common sense dictates usage, depending on location and need for water. A well yield of 2 gallons per minute (0.13 L/s) would define an aquifer for a rural household that needs a home water supply, but it wouldn't define an aquifer for an engineer looking for a town water supply.

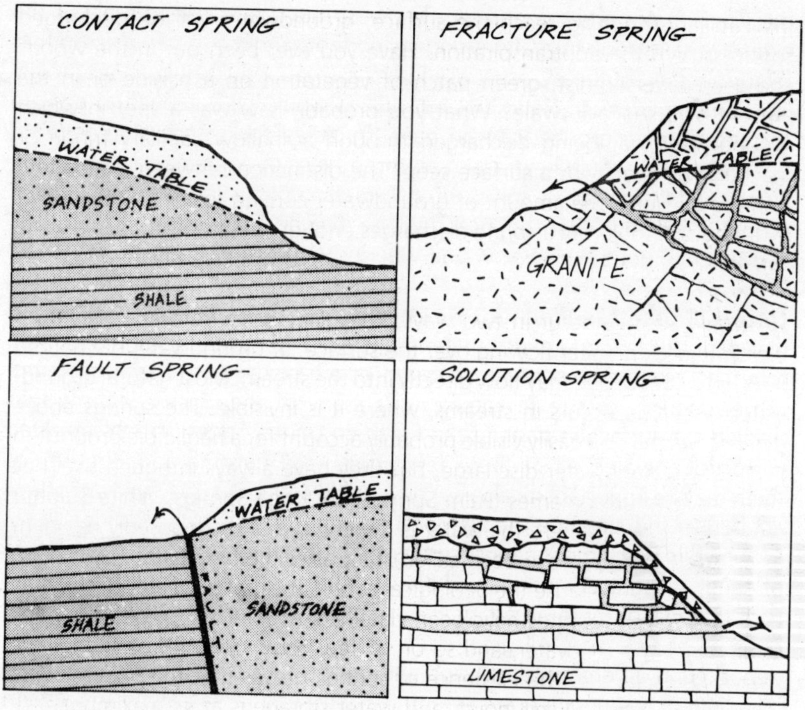

FIGURE 7.2 Common types of springs.

Geology of Aquifers

Although this chapter emphasizes the "natural" occurrence of groundwater, the subject is hard to study without some human intervention. While aquifers may crop out on the surface in areas of recharge and discharge, the actual path of groundwater flow occurs underground where we can't see it. Thus, the study of aquifers requires drilling wells to outline the extent and hydrologic nature of the water-bearing formations. As will become obvious, much of what is known about aquifers is derived from well information.

If you asked well drillers what kinds of geologic formations they consider to be aquifers, answers would probably vary depending on where the drillers lived. Drillers everywhere would agree that unconsolidated sand and gravel often provide good aquifers. Beyond that, however, you would notice distinct regional preferences for other kinds of rocks. Drillers in the Pacific Northwest would tell you about the wells they developed in volcanic rocks. Those in the Plains States and the northern Midwest might talk at length about the extensive sandstone formations underlying thousands of square miles in that region. In the Southeast, especially in Florida, you would hear about prolific limestone aquifers. In New England, drillers would be equipped to explore for and develop groundwater in hard, fractured rocks such as granite. In California, you would hear about all these types of aquifers, except for the limestone. After such a survey of drillers around the country, you would probably conclude that the local geology may be as important as the local hydrology in determining how and where groundwater occurs.

Unconsolidated Sediments. An unconsolidated sediment is a loose, granular deposit of natural earth materials in which particles are not cemented together. Most soils are unconsolidated, as are sands at the beach, sand and gravel deposits of stream channels, or many of the glacial deposits that blanket the surface in parts of the northern United States. When saturated with groundwater, all of those formations may become aquifers. Any of them may be important locally, but on a worldwide basis, the most important aquifers in this class are the alluvial sediments.

Alluvium is sediment deposited by running water, consisting of a mixture of sand, silt, gravel, and clay. These are composed of siliceous mineral or rock material of more or less similar composition. They are classified on the basis of particle size. All grain sizes are carried by running water but are deposited in different places. Sand and gravel, carried along the stream bed by swiftly flowing water, are eventually deposited in and along the stream channels. Finer-grained silts and clays are often carried in suspension in the moving water and are more likely to be deposited when water overflows the banks and spreads out on floodplains adjacent to the streams. In time, alluvial deposits may accumulate to form a more or less random pattern of coarse-grained channel deposits enclosed in finer-grained floodplain deposits (see Figure 7.3). When saturated with groundwater, the whole mass of alluvium (except for pure clay bodies) will yield water to wells, but the coarse-grained deposits make the most productive aquifers. Most alluvial aquifers contain relatively young, unconsolidated sediments.

Consolidated Sediments. Unlike alluvium, which generally occurs as local deposits of restricted areal extent along stream valleys, many consolidated sediments are areally extensive, sometimes forming major landscape features. For example, in the U.S. Southwest, consolidated sediments have been eroded to form spectacular scenery in such places as Zion, Bryce, and Grand Canyon National Parks. When they occur in the subsurface, as they do beneath much of the Great Plains, consolidated sediments are important regional aquifers.

Individual beds have a mineral composition similar to the unconsolidated sediments, but the originally loose grains are now compacted and cemented

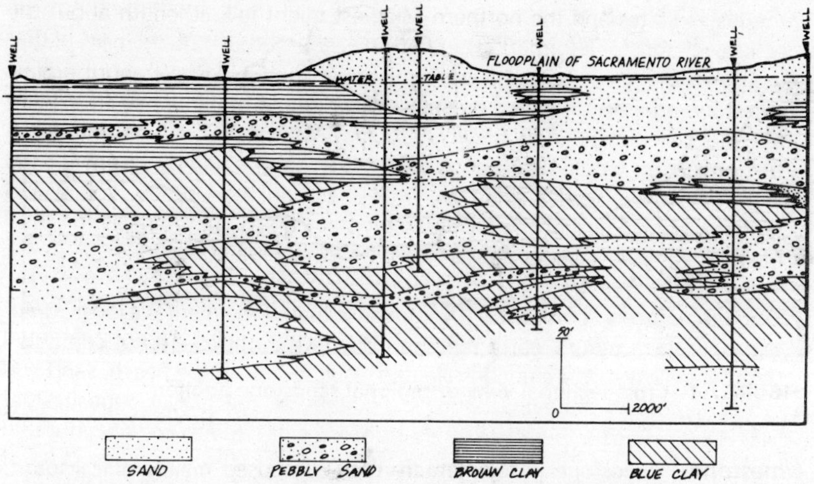

FIGURE 7.3 Cross-sectional view of alluvial sediments just west of Sacramento, California.

together. They have become sedimentary rocks. Sand has changed to sandstone, silt to siltstone, and clay to shale. Limestone is also a sedimentary rock, but its special hydrologic character places it in a separate category.

Although several types of sedimentary rocks may be important locally as aquifers (e.g., coal in the northern Plains States), sandstone is the most important throughout the world. It accounts for about 25% of all sedimentary rocks, and in many parts of the world, sandstone beds form extensive groundwater reservoirs. Sandstone grains are mostly silicate minerals, which are relatively insoluble in water. A typical basin with a sandstone aquifer is shown in Figure 7.4. Sandstones, as compared with alluvium, tend to exhibit more uniform hydrologic properties. The coarse, well-sorted sand fraction of alluvium may have 40% or more porosity and will be very permeable to water flow. In a section of alluvial aquifers (as in Figure 7.3), though, porosity and permeability may vary widely. Sandstone, depending on degree of compaction, amount of cement, and so on, may range in porosity from 2% or 3% up to about 30%, with a corresponding range of permeabilities. Porosity and permeability are generally fairly uniform throughout a particular sandstone bed. Because of its broad areal extent, sandstone frequently stores very large volumes of water. An example is the sandstone that supplies water to the town of Alice Springs in central Australia. This formation, more than 1000 feet (300 m) thick and underlying several thousand square miles south and west of Alice, has been estimated to contain water to supply the town for hundreds of years.

FIGURE 7.4 Cross-sectional view of regional sandstone aquifer.

Limestone. Limestone is sedimentary rock composed mainly of calcium carbonate ($CaCO_3$), formed by precipitation of $CaCO_3$ from natural waters or by organic processes involving lime-secreting animals such as plankton, algae, corals, and shellfish. Limey deposits often have high original porosities, but as they are changed to sedimentary rocks through compaction (due to burial beneath younger sediments), they often become dense, hard rocks with lower porosity. Sometimes the high temperatures and pressures due to deep burial change the rock to marble. Although geologists classify marble as a metamorphic rock because of its mode of formation, it is still limestone as far as hydrologic processes are concerned.

Limestone owes its unique hydrologic properties to the solubility of its major constituent, $CaCO_3$, in water. Rainwater, seeping down into the cracks of a limestone rock, nearly always carries carbon dioxide (CO_2) in solution. This gives the water a slight acidity and causes a certain amount of $CaCO_3$ to dissolve. While you wouldn't detect much enlargement of the limestone cracks in a month or even in a year, over a few hundred years the openings would widen noticeably. The ultimate result could be a limestone cavern, but generally it becomes just a slightly enlarged solution opening that allows water to infiltrate and flow through the rock. Eventually, even the densest limestones can acquire enough porosity and permeability to become good aquifers (see Figure 7.5).

Volcanic Rocks. Volcanic rocks, especially lavas, make good aquifers in many parts of the world. They are important as water-bearing formations because of their widespread occurrence and because of their distinctive systems of fractures and porous zones. The most abundant and widespread volcanic rocks are the basalt lavas that make up many volcanic islands in the sea (e.g., the Hawaiian Islands) and that form the great lava plateaus such

FIGURE 7.5 **Limestone aquifer showing groundwater in solution openings.**

as the Deccan Plateau in India and the Columbia Plateau in the northwestern United States.

Basalt lava, where it covers extensive areas, often provides both conduits for groundwater transmission and reservoirs for water storage. The vertical fractures in many basalt flows provide pathways for rainwater to penetrate the otherwise dense lava rock and seep down to porous zones between flows (Figure 7.6). Shrinkage during cooling of the lava causes the fractures, which give outcrops of basalt their *columnar* appearance. The porous zones between flows sometimes result from a blocky, broken crust at the top of

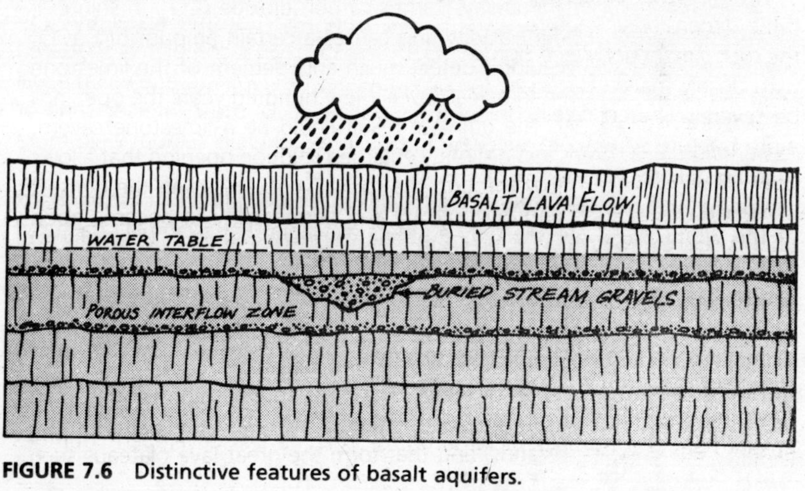

FIGURE 7.6 **Distinctive features of basalt aquifers.**

an underlying flow, and sometimes from porous sediments on top of older flows deposited between eruptive periods.

Porous interflow zones are often efficient, long-range conduits for groundwater. For example, in southern Idaho on the Snake River Lava Plains (a part of the Columbia Plateau Province), an ancient drainage system on the north side of the Snake River was flooded by lava and buried beneath hundreds of feet of volcanic flows many thousands of years ago. Here the rivers draining the Lost River range empty onto the lava plain and disappear, only to reappear far to the south as large springs in the walls of the Snake River gorge. The combination of porous alluvium in the buried river channels and the interflow zones in the lava provides an efficient conduit for transporting very large volumes of groundwater over long distances.

Crystalline Rocks. As used here, the term *crystalline* refers to what geologists call igneous and metamorphic rocks. They come in many varieties and have many names. Granite is the most common crystalline igneous rock; slate is a common metamorphic rock. Rock names and formation are not important, except as they influence the rocks' hydrologic properties. Volcanic rocks are also igneous rocks, but because they have very different hydrologic properties, they are considered in a separate category.

Crystalline rocks share two important attributes: (1) they are almost always uniformly hard, dense rocks; and (2) they have very low porosities and permeabilities. Porosity in an unweathered, unfractured rock might range between 1%–3% and will commonly be below 1%. Permeabilities are also correspondingly low. Since aquifers depend on porosity for storage and permeability for water transmission, crystalline rocks generally are poor aquifers.

Groundwater reservoirs in these rocks appear mainly in interconnected, three-dimensional fracture systems that store and transmit water through the rock's main body (see Figure 7.7). Unless the rock is completely shattered with closely spaced fractures, the groundwater reservoir storage volume will be severely limited. Wells drilled in crystalline rocks must intersect one or more fracture systems to succeed. Even successful wells seldom yield more than 50 gallons per minute (3 L/s).

Why would anyone bother to drill wells in crystalline rock? In some parts of the world these are the only aquifers available. For example, fractured rock underlies about half of the Australian landscape. Where rain or snowmelt is sufficient to recharge the groundwater, fracture reservoirs in these hard rocks may provide a satisfactory supply for individual domestic water systems. In the Republic of South Africa, for example, many farms rely on wells tapping fractured rock, but only about 10% of South Africa's water supply comes from the ground. Most of the population uses surface water stored in reservoirs behind a number of large dams. In some less-developed countries

FIGURE 7.7 Groundwater in fractured granite.

that have insufficient capital to build dams or few suitable surface reservoir sites (e.g., parts of India), groundwater from fracture reservoirs may be the only perennial supply. These places often have chronic water shortages.

Confined and Unconfined Aquifers

Regardless of the kinds of rocks an aquifer comprises, groundwater may occur in two fundamentally different ways in nature: (1) unconfined beneath a free water table; or (2) confined beneath an essentially impermeable formation that seals off the aquifer from above. The water table is defined as ". . . that surface in the groundwater body at which the water pressure is atmospheric." *Unconfined* groundwater, then, is water in an aquifer that has a water table in contact with the atmosphere through pores in the unsaturated soil above. Unconfined aquifers are sometimes called *water-table* aquifers.

Confined groundwater, on the other hand, is water under pressure greater than atmospheric pressure. The upper boundary of a confined aquifer is an essentially impermeable formation that "traps" or "confines" water in the aquifer, sealing it off from the atmosphere (see Figure 7.8).

From a practical standpoint, how does one distinguish between confined and unconfined aquifers in the field? The United States Geological Survey uses these criteria:

1. An unconfined aquifer has a water table, which is defined by the levels at which water stands in wells that penetrate the water body just far enough to hold standing water. This means that when a water-

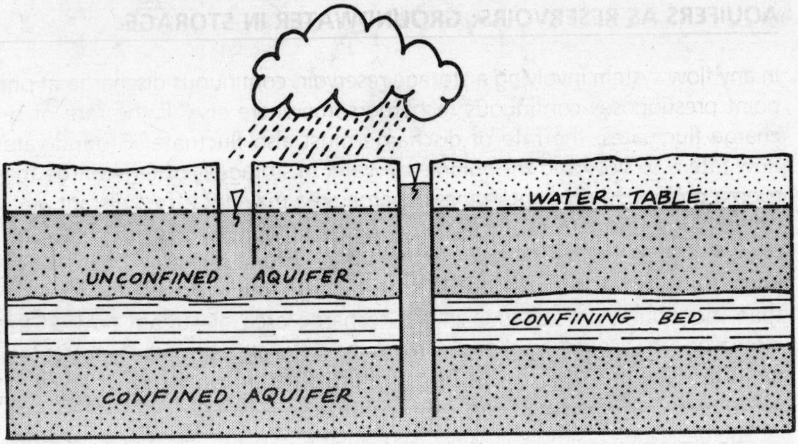

FIGURE 7.8 Typical occurrence of groundwater in confined and unconfined aquifers.

table well is drilled, water will remain approximately at the same level where it is first encountered.

2. In wells tapping a confined aquifer, the water will rise in the well when it is first encountered during drilling, and will stand at a level above the top of the aquifer.

Water in a confined aquifer is under pressure because the aquifer is at a higher elevation in the recharge area than it is at the well location, just as water in your kitchen faucet is under pressure because the neighborhood storage tank is at a higher elevation. Depending on local conditions, water from a confined aquifer will even rise up in the well until it flows out at the surface without the aid of a pump. Traditionally such a well is called an *artesian* well, after the French city of Artois, the site of an old Roman town called Artesium famous for its flowing wells during the Middle Ages. Confined aquifers are often called *artesian* aquifers.

The Function of Aquifers in the Water Cycle

Aquifers have two main functions in the underground phase of the water cycle. They store water for varying periods in the underground reservoir, and they act as pathways or conduits to pass water along through the reservoir. Although some are more efficient as pipelines (e.g., cavernous limestones) and some are more effective as storage reservoirs (e.g., sandstones), most aquifers perform both functions continuously.

AQUIFERS AS RESERVOIRS: GROUNDWATER IN STORAGE

In any flow system involving a storage reservoir, continuous discharge at one point presupposes continuous recharge somewhere else. If the rate of recharge fluctuates, the rate of discharge must also fluctuate. Groundwater flow rate is so slow and the volume of water in storage is often so large that even a prolonged drought in the recharge area may not affect distant areas of discharge for many years. Likewise, heavy pumping from wells may continue for years before the effects reach the natural recharge or discharge areas. Keeping track of conditions in the underground reservoir requires more than monitoring precipitation in the recharge area or stream flow in the discharge area. Detecting changes in groundwater storage means using other methods. The problem can be approached in several ways, all of them involving periodic measurements of groundwater levels.

The basic idea is simple. In any reservoir storing water or other fluid, rising fluid levels indicate an increase in storage and falling levels indicate a decrease. With the dimensions of a surface reservoir known, calculating exactly how much the storage will change for each foot or meter change in the water level is simple.

The situation underground is a bit more complicated because the storage space, which is related to aquifer porosity, is only a fraction of the aquifer's total volume. Multiplying the linear change in water level by a fractional coefficient yields the actual change in groundwater storage. This multiplier, the *storage coefficient,* is defined as the volume of water going into or coming out of storage per unit surface area of the aquifer per unit change in hydraulic head, where hydraulic head is the height of water in the well above a standard datum (such as sea level). This coefficient is the ratio of active storage space to total aquifer volume and is therefore a dimensionless number. For water-table aquifers the storage coefficient is approximately equal to the specific yield, as defined in Figure 5.5. For an artesian aquifer the concept of storage coefficient has a different physical meaning, as will be explained shortly.

Storage in Unconfined Aquifers

To understand why underground water levels must be multiplied by a fractional coefficient to obtain storage volumes, it might be instructive to consider some similarities and differences between surface reservoirs and water-table aquifers. Many surface reservoirs have two storage components—live and dead. Dead storage is the volume of water below the intake tower or penstock inlet behind a dam, or the water in a tank bottom below the outlet level. For all practical purposes this water is unusable. Live storage is the remainder of reservoir volume; it contains water that can be withdrawn from the reservoir for use. Changes in live storage are easily calculated using the reservoir dimensions and the amount of increase or decrease in the water level. Usually a sharp line of demarcation separates the two storage units, each occupying a distinct volume in the reservoir that is easy to calculate separately.

As with a surface reservoir, an unconfined aquifer's pore space also contains both live storage and dead storage. *Gravity water* makes up the live storage and *capillary water* the dead storage. When the water table rises, it saturates the overlying capillary fringe; when it falls, it leaves capillary water suspended above it. In either case only gravity water actually moves into or leaves storage in the aquifer, causing water levels in wells to rise or fall. For an unconfined aquifer, the ratio of aquifer volume holding capillary water is called *specific retention;* the ratio of aquifer volume containing gravity water that is free to move in or out of the pores is called *specific yield.* Specific retention plus specific yield equals porosity, which is the ratio of total open space to total volume of the aquifer. Referring again to Figure 5.5 may help to clarify those relationships.

Once the fine-grained pores in an aquifer acquire their natural quota of capillary water (specific retention), they never again release it to gravity drainage. Water levels in wells do not indicate how much water is held in this volume of dead storage. *Specific retention* for groundwater is similar to *field capacity* for soil water; both terms concern the amount of pore water held by capillary forces against gravity. No term in soil science corresponds to specific yield, except perhaps "deep percolation." Soil scientists often go to considerable lengths to measure field capacity, the retained water, and less so trying to measure the deep percolation that drains away below the root zone. Groundwater hydrologists, on the other hand, try to estimate specific yield and don't worry much about specific retention. The two emphases reflect their interests: Soil scientists want to know how much water can be put *into* the ground to grow a crop, while groundwater hydrologists want to know how much water they can get *out* of the ground from wells.

Storage in Confined Aquifers

The implications for changes in storage as water levels fluctuate are very different for confined versus unconfined aquifers. In an unconfined aquifer, lowering the water table actually dewaters the upper part of the aquifer and reduces the volume of saturated material, but the aquifer and the water undergo essentially no physical changes. When water levels fall in wells tapping a confined aquifer, the pressure falls, but the aquifer remains completely saturated. The change in pressure in a confined aquifer affects both the water and the aquifer.

The aquifer responds to pressure change by expanding or contracting slightly as water levels rise or lower. This produces a small increase or decrease in porosity, and hence in storage space. Water, although usually thought of as being noncompressible, will also contract or expand slightly in response to pressure changes. Thus, when water is removed from storage, well levels fall, pressure in the aquifer falls, water expands slightly, and the aquifer contracts slightly. When water is added to storage, the reverse processes take place: well levels rise, pressure in the aquifer rises, water contracts slightly, and the aquifer expands slightly. Because the change in aquifer porosity or in water volume per unit change in pressure is very small, the numerical value of the storage coefficient for an artesian aquifer is much smaller than that for a water-table aquifer. The range in values for the storage coefficient in water-table aquifers is around 0.01–0.30, while the numbers for artesian aquifers are more like 0.00005–0.005.

It is now clear why water levels in wells tell quite a different story regarding groundwater storage, depending on whether the wells tap confined or unconfined aquifers. For a well tapping an unconfined aquifer, a drop of 1 foot (30 cm) in water level means the dewatering of that much formation and may represent a considerable volume of water. When the water level falls 1 foot (30 cm) in an artesian well, it means a slight reduction in pressure, but a much smaller volume of water removed from storage.

Storage in Relation to Geology

Measuring water levels in wells helps to determine what the water is doing underground, but that alone doesn't tell you very much about the three-dimensional structure of the storage reservoir. A realistic interpretation of water levels requires at least some knowledge of the configuration and internal composition of the underground formations containing groundwater. How thick is the aquifer? What is the nature and composition of the aquifer and any confining beds? If the well water is of poor quality, what is its source? How thick is the total producing zone (how deep is the effective bottom of the reservoir)?

Some of this knowledge can be gained by geologic mapping in areas where the aquifers crop out on the surface; however, not all aquifers appear on the surface, and even when they do, the outcrops may differ considerably from the subsurface part of the formations that are penetrated by wells. To study these buried reservoirs, the groundwater hydrologist uses a technique called *subsurface geology.*

Subsurface geology, originally developed by petroleum geologists, involves the same things that concern the surface geologist. Instead of observing the formations directly, though, the subsurface geologist can characterize rock bodies by inferring their composition and properties from rock samples collected during drilling and from geophysical measurements made in the holes at the completion of drilling. The result is a vertical section of the subsurface showing the kinds of rocks and their relationships to each other. A typical section, interpreted from well data, appears in Figure 7.3, which also shows the well control used in drawing the subsurface picture. As you might suppose, the subsurface geologist must have a good imagination as well as considerable geological skill and must be ready to alter the picture whenever a new well is drilled. On the positive side, however, a good subsurface section can be useful in helping to interpret water levels correctly and in planning future wells. As you may have guessed, most groundwater hydrologists start their careers as geologists.

INTERPRETATION OF WATER LEVELS IN A CALIFORNIA RESERVOIR

Perhaps the best way to show what water-level measurements can indicate is to consider some real-world examples. The San Joaquin Valley of California (see map in Appendix I) is one of the most intensively developed groundwater basins in the world. Here approximately 50,000 wells pump about one-fourth of the groundwater used for irrigation in the United States. Dozens of alluvial aquifers, deposited by rivers draining the adjacent mountain ranges in late geologic time, comprise the groundwater reservoir. Ancient lake beds of clay and silt also separate unconfined from confined water at various places throughout the valley. The main underground reservoir, which is about 300–3,000 feet (100–900 m) in depth, is simple in general outline but complex in local detail. Its magnitude can be gaged according to storage capacity. The United States Geological Survey has estimated that the valley as a whole has space for about 93 million acre-feet (115,000,000,000 m³) of groundwater storage in the first 200 feet (61 m) of sediments below the surface. The area considered here is at the southern end of the valley, location of the most intensive development of groundwater. While several deep, confined aquifers are located in the region, the examples shown here are all from the larger, more extensive water-table aquifers that contain the near-surface groundwater (see Figure 7.9).

Declining Water Levels and Storage Reduction

Historically nearly all groundwater recharge came from the Kern River, which drains a large area of snowfields in the Sierra Nevada. For most of its course across the valley floor the Kern loses water, which infiltrates through the stream bed and drains down into the underlying groundwater. This makes a groundwater mound in the subsurface. In recent years two large canals have been built to bring additional irrigation water into the region from the north. The southern valley is almost a closed basin, and before the wells

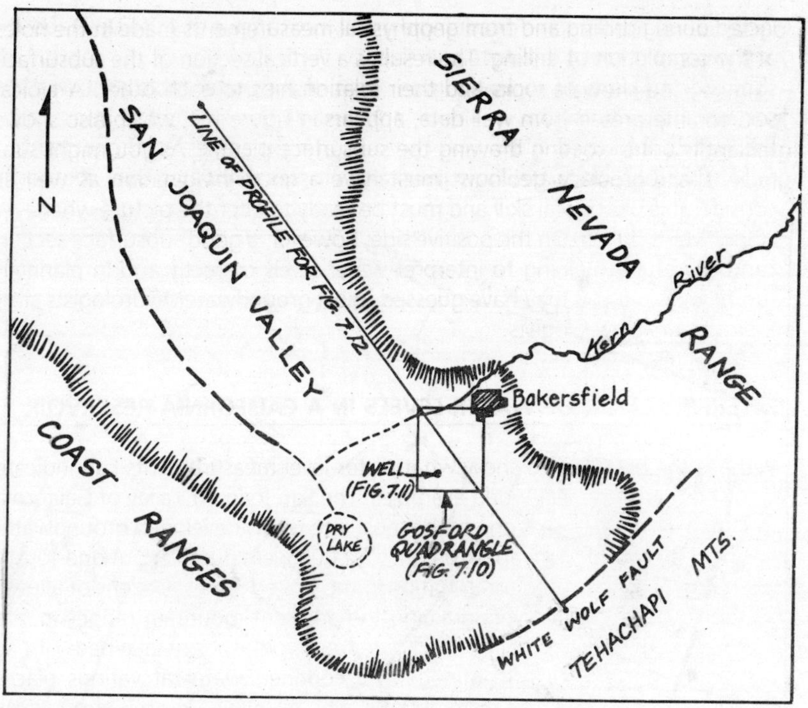

FIGURE 7.9 Southern San Joaquin Valley, California.

were drilled most natural discharge of groundwater occurred through evapotranspiration from swamps and phreatophytes on the valley floor. After nearly a century of pumping for irrigation, most groundwater discharge now occurs through wells, and the regional water table has been lowered many feet. Federal, state, and local agencies have measured groundwater levels for most of the century, providing a large volume of data for study. The following examples were plotted using data from these sources.

Water-Level Contour Map

Using water-level elevations in wells as control points, a contour map of the water table can be drawn in the same way a topographic contour map is drawn using land-surface elevation. In every year since 1920, the State of California has published a generalized contour map of groundwater levels for the San Joaquin Valley as a whole. These small-scale, regional maps are valuable sources for groundwater hydrologists making regional studies. Detailed studies of local areas often require constructing maps on a larger scale. The base maps most commonly used are the standard United States Geological Survey topographic quadrangle maps, with a scale of 1:24,000 (about 7 × 9 miles or 11 × 14 kilometers in size).

Figure 7.10 shows a typical contour map of the water table as it was in 1976 for the area shown in Figure 7.9. Just as the steepest topographic gradient is a line crossing the land contours at right angles, so the hydraulic gradient (and groundwater flow lines) also crosses the water-level contours

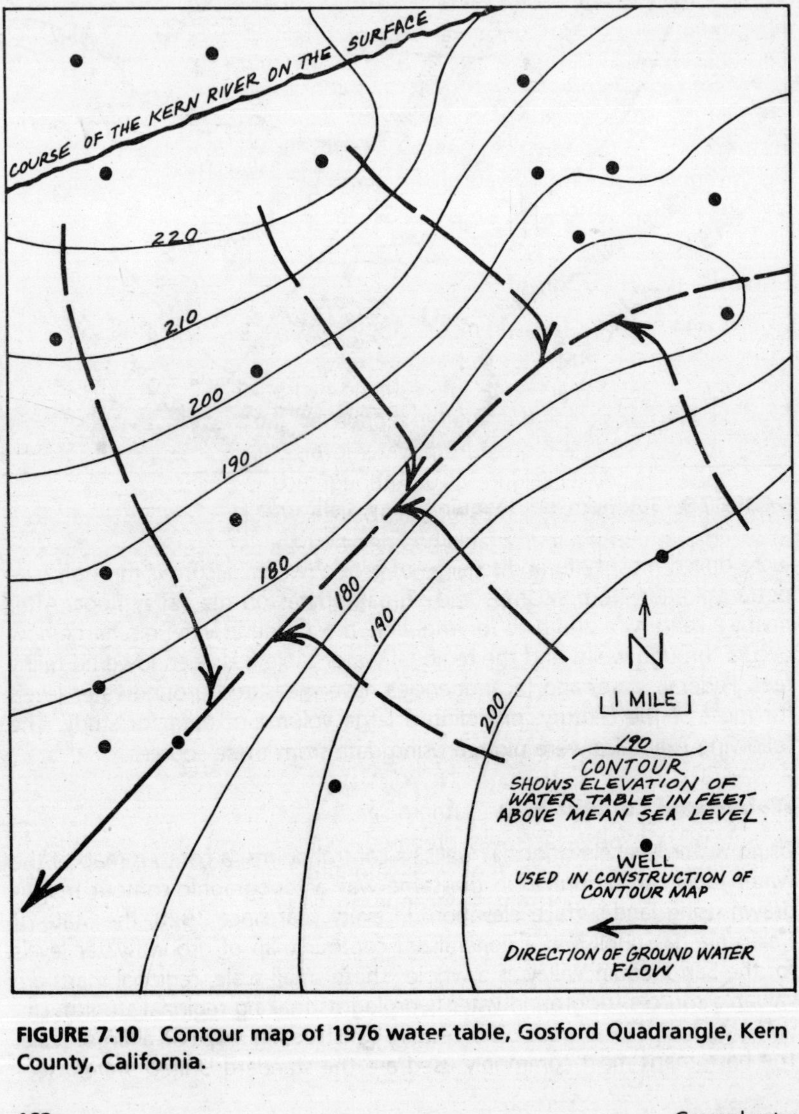

FIGURE 7.10 Contour map of 1976 water table, Gosford Quadrangle, Kern County, California.

at right angles, shown on the map by dotted lines. One use of such a map is to show direction of the groundwater movement in the aquifer. It also indicates how far the water table is below ground surface so land owners and well drillers will know at least the minimum depth to plan for new wells. The map's greatest use, though, is as evidence of groundwater storage changes. A series of maps made over a period of years helps to visualize in three dimensions changes in the groundwater reservoir. For example, in the map area shown in Figure 7.10 the 1925 water table averaged about 10 feet (3 m) below ground surface; by 1976, that level had dropped to about 170 feet (52 m). How did such a decline occur? Preparing 50 contour maps for the 1925–76 period would be tedious, and there is an easier way to get this information: A groundwater hydrograph will show the timing and rate of water-level recession.

Groundwater Hydrograph

A series of water-level measurements plotted against time is called a groundwater hydrograph. Annual measurements are usually taken from a single well over a span of years, with the resulting hydrograph showing the long-term storage trends for the aquifer tapped by the well. The hydrograph shown in Figure 7.11 is plotted from water levels measured in a well located within the area shown in Figure 7.10. Although no measurements after 1969 are available for this well, measurements in a well about a mile (1.6 km) away indicate that the water table had declined about another 20 feet (6 m) by 1976. Clearly, discharge of groundwater through wells was much greater than recharge from the Kern River during the period represented in the hydrograph.

Water-Level Profile

The water-level contour map and hydrograph show action of the water table in a restricted area. To indicate the regional effect of the long-term decline in groundwater storage in the southern San Joaquin Valley, a profile (or cross section) of the subsurface was prepared. Figure 7.12 shows this profile, plotting the ground surface, the 1925 water table, and the 1976 water table on the same scale. Approximate location of the line of profile is shown on the map in Figure 7.9.

Using records of river flow and canal-water imports, and estimating total water losses due to evapotranspiration from the large acreage of irrigated crops, the net extraction of groundwater from this reservoir appears to be about 25 million acre-feet (31,000,000,000 m^3) for the 1925–76 period. The size of this natural resource is indicated by the fact that, even though the basin is severely overdrafted now, much groundwater remains in storage.

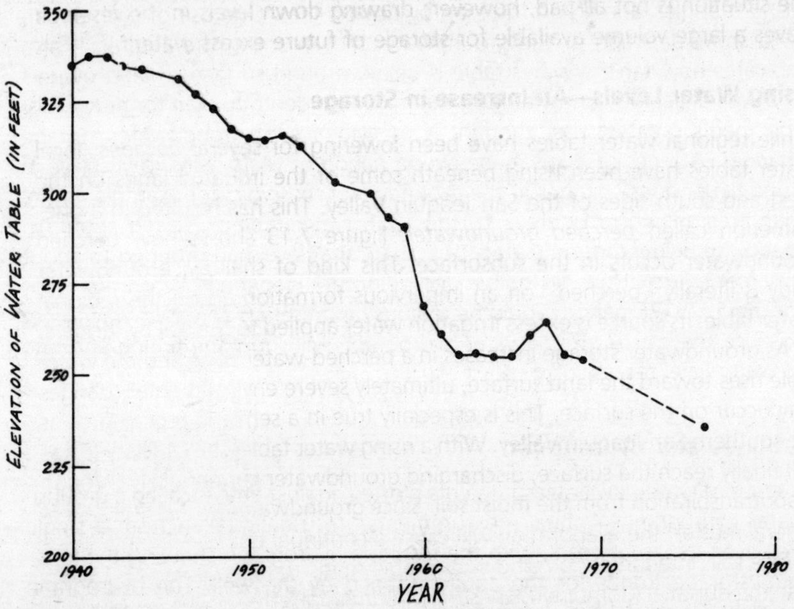

FIGURE 7.11 Groundwater hydrograph (See Figure 7.9 for location of well).

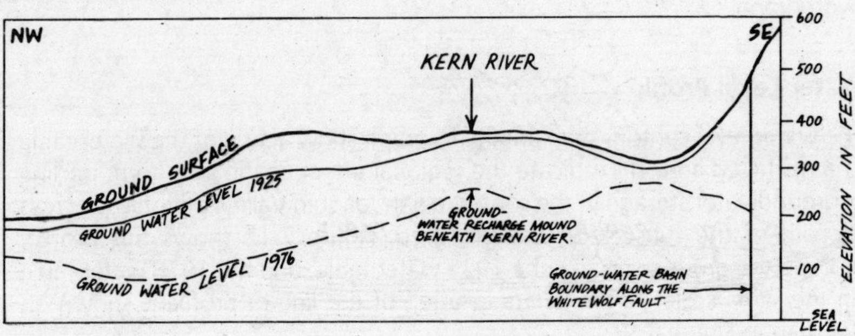

FIGURE 7.12 Cross-sectional view of groundwater reservoir in southern San Joaquin Valley, California. Profile is oriented NW–SE and extends from the south for about 60 miles (97 km) northwesterly along the middle of the valley. Approximate location is shown on the index map. Shown on the profile are the 1925 and 1976 average groundwater levels.

The situation is not all bad, however; drawing down levels in the reservoir leaves a large volume available for storage of future excess water.

Rising Water Levels—An Increase in Storage

While regional water tables have been lowering for several decades, local water tables have been rising beneath some of the irrigated lands on the west and south sides of the San Joaquin Valley. This has resulted in a phenomenon called *perched groundwater*. Figure 7.13 shows how perched groundwater occurs in the subsurface. This kind of shallow, groundwater body is literally "perched" on an impervious formation above the regional water table. Its source is excess irrigation water applied to the soil by farmers.

As groundwater storage increases in a perched-water zone and the water table rises toward the land surface, ultimately severe environmental changes may occur on the surface. This is especially true in a semiarid region such as the southern San Joaquin Valley. With a rising water table, the capillary fringe will finally reach the surface, discharging groundwater continuously through evapotranspiration from the moist soil. Since groundwater contains dissolved mineral matter, the evaporation will cause a continual buildup of salts in the surface soil. In addition, a rising water table tends to drown out plant roots near the surface. Combined, these environmental effects spell trouble for farmers, even though groundwater storage is increasing. This problem has plagued irrigation agriculture in arid lands for thousands of years, since the days of the Hittites and the Babylonians.

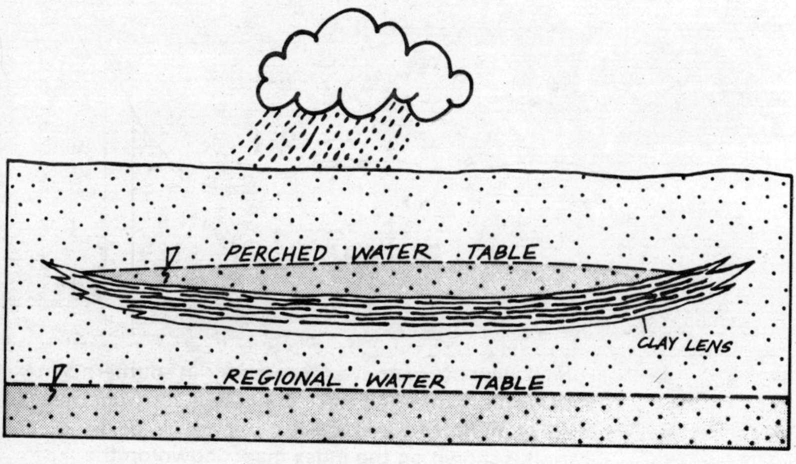

FIGURE 7.13 Perched groundwater.

A.

B.

FIGURE 7.14 (a) Test pit showing perched water table. Scale on paper hanging down into pit is in feet. Regional Water Table is more than 100 feet below surface at this location. (b) Soil dug from pit showing white crusts of salt left in soil by evaporation of water from capillary fringe above perched water table. Location is in southern San Joaquin Valley, California.

In the autumn of 1977, soil scientists and farmers met in the southern San Joaquin Valley to discuss the problem of perched groundwater and to suggest measures to combat its deleterious effects on agricultural crops. Several test pits were dug to illustrate the natural subsurface conditions (see Figure 7.14). In this fine-grained soil, the capillary fringe above the perched water table is practically at the ground surface, and the salt (the white crusts shown in the photo) was deposited by capillary water evaporating from the soil surface.

To counter the deleterious effects of this perched groundwater, farmers have joined to form drainage districts for draining the affected lands and for disposing of the mineralized drainage water. Ultimately the state may build a drainage canal to carry all valley drain water to the sea; meanwhile, the local drainage districts are using evaporation ponds to dispose of the unwanted water.

These examples from California not only illustrate some aspects of the reservoir function of aquifers, but also show how our intervention in the natural working of the water cycle can cause unforeseen environmental changes. It is ironic, for example, that much of the perched groundwater originated in deep irrigation wells that were depleting storage in the regional groundwater body while at the same time the shallow, unwanted perched water was rising toward the root zone. Lowering the regional water table by excessive pumping has also cut off the avenues for widespread, natural groundwater discharge through phreatophytes and evaporation from swampy ground. Now essentially no natural discharge comes from the main groundwater reservoir. Where irrigated crops have not replaced the original vegetation, the valley floor has become a desert. These examples are cited not to condemn civilization for changing the natural order of things, but merely to show that when it does, it must balance environmental costs against the benefits of water development.

AQUIFERS AS CONDUITS—GROUNDWATER IN MOTION

For a drop of water moving through the water cycle, the slowest leg of the journey is the one through the groundwater reservoir. In a race between a slow tortoise on the surface and a drop of water underground, the tortoise would win almost every time. Flow in surface streams is usually expressed in feet or meters per second, while groundwater flow typically might be a fraction of an inch (a few millimeters) to a few feet or meters per day.

Exceptional cases of rapid flow have been reported in open fissures of volcanic rocks or in solution openings in limestone aquifers; here, flow may be as rapid underground as it is on the surface. Such conditions are rare, but they do exist and have given rise to the misconception that groundwater flows in underground "streams." While water may indeed run in an underground stream for a short distance in a few places, in almost all cases groundwater "seeps" or "percolates" very slowly through tiny pores and cracks in the aquifers that make up the underground reservoir.

In one typical sandstone aquifer of average permeability in the Southwest, the United States Geological Survey found the rate of groundwater movement to be about one foot (30 cm) per week under a hydraulic gradient of 10 feet per mile (1.9 m/km). This aquifer is about 200 feet (61 m) thick, and along a 60-mile (97 km) cross section, where it is tapped by many irrigation wells, it provides a computed flow of about 24 million gallons (91 million L) per day. Conditions of flow in an aquifer like this one are obviously much different from those controlling flow in a surface stream. Dealing with groundwater requires a different perspective from the one most people have when they think of running water.

Sometimes it is pleasant to sit idly on the bank of a stream and watch the water flowing by. Probably everyone reading this book has done that at one time or other; the mere suggestion brings to mind a visual image of running water. Imagine now, if you will, that you are sitting on the "bank" of the aquifer described above. Can you picture water flowing by at a foot a week? Hardly! Even in one's imagination it is practically impossible to visualize movement so slow. In comparison, the hour hand of a clock literally races around the dial. Yet the groundwater must be moving, because the wells are producing year after year, and the aquifer remains full. This is indeed an enigma. Flow conditions underground must be radically different from those on the surface.

The most sluggish creek or river moves many times faster than a foot a week, even on an essentially level surface. So, knowing that water obeys the same laws of physics underground as on the surface, one would suspect that the slope, or gradient, down which water flows in an aquifer may differ from the slope of a stream bed on the surface. Also, if a surface stream can't possibly flow as slowly as most groundwater flows, then the channels through which groundwater moves must be radically different from stream channels, too. How does the gradient of a surface stream differ from the gradient down which groundwater moves? What kinds of channels would cause water to creep along so slowly through the aquifer? When we have answered these questions, we will have gone a long way toward understanding underground flow in the saturated zone.

Water running in a stream continually loses some of its potential energy as it moves down the stream bed slope, since constant conversion of potential energy (energy of position) to kinetic energy (energy of motion) propels the moving stream. The topographic gradient of a stream bed is a gravitational energy gradient as well.

In general physical terms a *gradient* expresses the change of one variable with respect to another variable. For example, the geothermal gradient (at a specific location) expresses the increase in temperature with depth below the surface. The topographic gradient, or grade, of an inclined land surface is expressed as the change in vertical distance with respect to horizontal distance—in other words, steepness. In Figure 7.15, the gradient of the hillside would be the ratio of cb/ba. If cb = 5 and ba = 100, the gradient of the inclined surface would be 5/100, or 0.05. By moving the decimal point two places to the right, the gradient can be expressed as a percentage. In this example the surface ascends at a 5% grade. This is a common way of expressing highway slopes.

Water flowing in an aquifer loses potential energy as it moves down a gradient, just as water in a stream does. In both cases water moves down energy gradients. The energy gradient underground is called the *hydraulic gradient;* it expresses flow in a unique environment very different from the land surface. The hydraulic gradient is determined by values of hydraulic head in the aquifer.

Hydraulic head (H) is defined as the elevation, with respect to a standard reference level, at which water stands in a well terminated at a specific point in an aquifer. For practical purposes the standard reference level is usually taken as mean sea level, which is the same reference used by most land surveyors and mapmakers (see Figure 7.16). The physical significance of hydraulic head lies in the fact that it is a single measurement representing the gravitational potential energy of water in the aquifer at that point.

Hydraulic gradient is defined as the decrease in hydraulic head per unit distance in the direction of flow. As shown in Figure 7.17, it is the difference in hydraulic head divided by the length of flow path between the wells where hydraulic head is measured. Whatever the inclination of the gradient, it always decreases from a higher to a lower elevation as potential energy is converted to kinetic energy along the flow path. This downward trend in

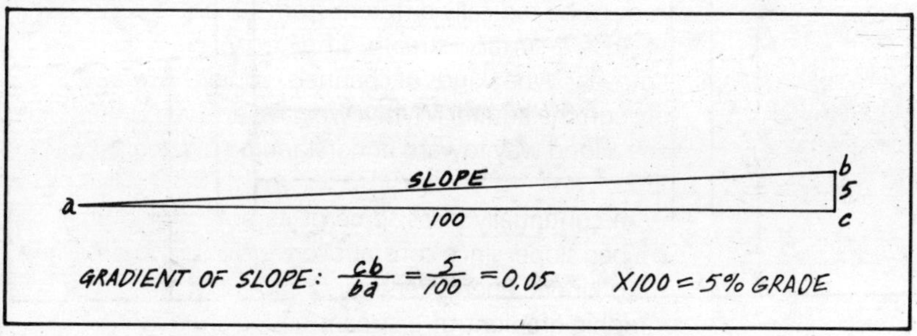

FIGURE 7.15 How gradient of a slope is determined.

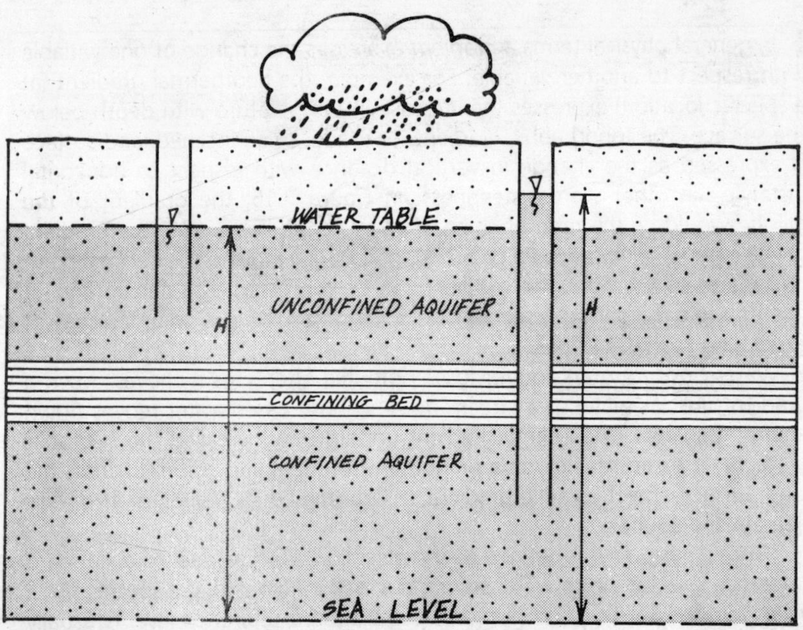

FIGURE 7.16 How hydraulic head (H) is measured.

the gradient reflects energy loss in the flowing stream due to viscosity (internal friction) of the water itself and to friction between the water and the pore channel walls. The same kind of energy loss occurs in pipelines (see Figure 7.18). This is the energy gradient responsible for groundwater flow. While it is similar to the energy gradient for surface flow, there is a subtle but important difference. Surface water always flows down a physical slope—a

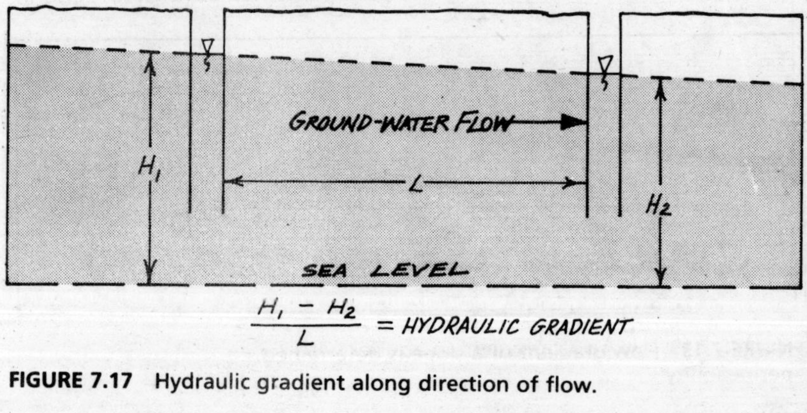

$$\frac{H_1 - H_2}{L} = \text{HYDRAULIC GRADIENT}$$

FIGURE 7.17 Hydraulic gradient along direction of flow.

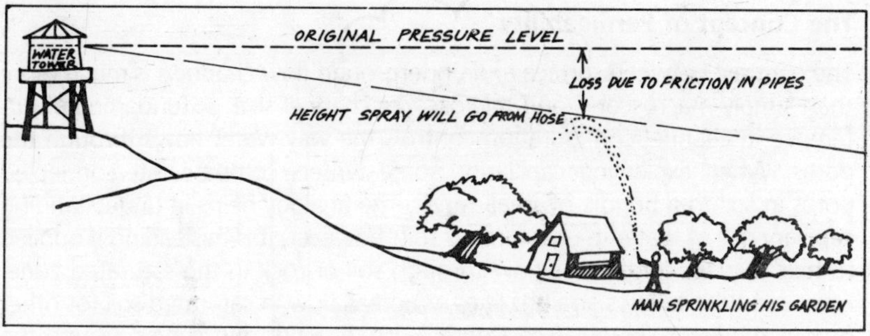

FIGURE 7.18 Head loss in water system due to friction in pipes (source: U.S. Geological Survey's "A Primer on Water").

stream bed, a hillside, an inclined driveway, and so on. Groundwater flows "down" an energy slope, but the flow channels themselves are not necessarily inclined downward in space (see Figure 7.19). The slope of the hydraulic gradient indicates the rate at which energy is used moving water through the aquifer pores. From Figure 7.19 it is clear that moving water through silt requires more energy than water moving through sand, even though both formations are inclined in space at about the same angle.

The flow paths in porous media are vastly different from the hills and valleys in a surface landscape. This difference in flow conditions is expressed in the concept of *permeability* as a distinctive property of porous media.

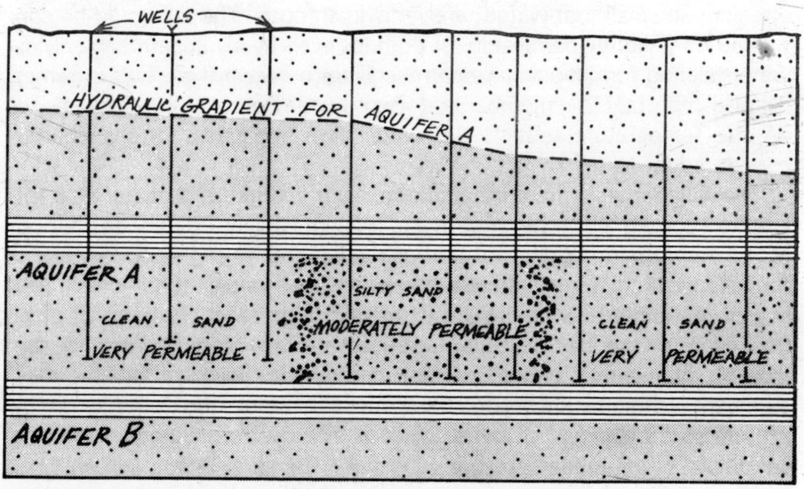

FIGURE 7.19 Hydraulic gradient in aquifer with zones of varying permeability.

The Concept of Permeability

The contrast between surface and underground flow channels is much more than simply macroscopic and microscopic channel size. A fundamental difference in channel configuration controls the way water flows through the pores. When explaining capillarity, some writers compare interconnected pores in soil to a bundle of small tubes. The analogy helps in understanding capillary rise of water in unsaturated soil; however, it is misleading if applied to help explain continuous flow through soil or rock in the saturated zone. Flow in a tube follows straight lines, whereas flow in saturated soil (or other porous media) is sinuous (see Figure 7.20). It winds and curves around the mineral grains, and it is much longer than a straight tube. As shown in Figure 7.20, the flow changes direction repeatedly, using some of the gravitational energy pushing the water along. Describing flow conditions accurately on a microscopic scale would be impossible, since flow occurs in each of the sinuous channels traversing a mass of porous material. Instead, the average flow conditions for all channels passing through a cross section of unit area at right angles to the flow path is summed up in the single term *permeability.*

Permeability is often spoken of as a porous medium's ability to transmit fluid. To be permeable, a porous material must have pores interconnected to provide pathways for fluid movement. The size and degree of continuity of the pore channels are the main determinants of permeability. If the pores are big and well connected, as in an open-work gravel, water will flow through freely and the gravel is said to be very permeable. A fine-grained formation, such as clayey silt, might have well-connected pores, but with openings so small that water barely moves through. The silt would be considered only slightly permeable, or perhaps even nearly impermeable. Most water-bearing formations have permeabilities between these two extremes. It is also important to remember that viscosity (see Figure 5.3), which depends on the temperature, affects the water flow regardless of the channel configuration.

The wide range in grain size and porosity of natural earth materials translates to a consequent wide range in aquifer permeability. Established methods of measuring permeabilities of porous media, both in the laboratory and in the field, result in a single number to express the complex conditions of flow in a given aquifer. Permeability is expressed using several systems of units, depending on the field of study—groundwater hydrology, soil science, petroleum engineering, and so on. The mathematics involved and the measurement techniques are beyond the scope of this book; this discussion simply introduces the concept of permeability as a single term summarizing flow conditions in an aquifer. Anyone who has watered a garden and watched water seep into soil has an intuitive understanding of this concept. Water

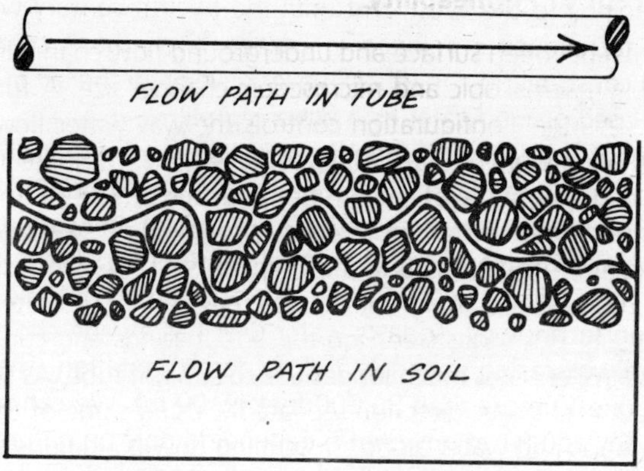

FIGURE 7.20 Flow paths for water in a tube and in soil.

disappears quickly into a loose, sandy soil, while a clay loam soil may accept water so slowly that water tends to pond on the surface. One is quite permeable; the other is less permeable.

Measuring aquifer permeability is useful in designing, testing, and pumping wells. Those who want to dig deeper into the subject will find a more technical discussion of permeability in several of the books listed in Appendix IV.[1]

Groundwater Discharge through Wells

As stated in the introduction to this chapter, humans have known how to dig wells since long before written records were kept. Some Egyptian oases used water from wells more than 4,000 years ago, and Moses probably learned the art of well digging from the Egyptians. Joseph's well, near the present city of Cairo, was dug about 1700 B.C. and extended to a depth of more than 300 feet (91 m). The ancient Chinese were even more advanced in the art of sinking wells and were the first to use a well-drilling machine and to use a well casing (bamboo) to keep the hole open. Some Chinese wells reportedly reached depths of several thousand feet. Unfortunately the lack of contact between the Chinese and Europeans prevented the transfer

[1]In the strict technical sense, *permeability* is a function of the porous medium alone. When the properties of the fluid are also considered, flow conditions in the porous medium are summarized by the term *hydraulic conductivity*. This technicality need not concern us here; however, in the event you pursue the subject further in some of the references listed in the appendix, you will need to learn definitions for both these terms.

of this technology to the West. Not until the twelfth century were modern well-drilling techniques developed in Europe. Even the Romans, first-rate engineers in constructing aqueducts, never developed the art of well drilling and seldom used groundwater as a water supply.

Settlement of arid regions during the past 150 years, especially in Australia and the western United States, has greatly stimulated groundwater development, causing major advances in the art of drilling and casing deep water wells. The parallel development of oil and gas reservoirs (since Drake's well at Oil City, Pennsylvania in 1859) has also promoted continuous advances in well-drilling technology. Today's water user has a wide choice of drilling methods and well-casing materials and no depth limitation. The deepest oil wells have gone to more than 30,000 feet (9100 m). We can now develop water from any aquifer at any depth, depending only on natural limitations of the aquifers, water quality, and, of course, money.

Why all this talk about wells in a book on the natural water cycle? If groundwater discharge through wells represented an insignificant proportion of aquifer discharge, it might be redundant (indeed, a century or two ago it would have been). At present, about 34% of all U.S. public water supplies, including one-half of our drinking water, comes from groundwater. This amounts to about 12 billion gallons per day (45 million m^3/day), a substantial amount even in terms of the water cycle.

Although rural areas still use hand-dug wells, especially in many less-developed countries, most wells today are drilled with well-drilling machines (called "drilling rigs" in the industry) and are cased with metal casing. Although water is lifted from dug wells in buckets or other containers in a few primitive societies, nearly all groundwater is lifted nowadays by mechanical pumps powered by either electrical or internal-combustion engines.

The Cone of Depression around a Pumping Well. The response of an aquifer to pumping is shown in Figure 7.21. When activated, the pump removes water from the well, with the head of water beneath the water table causing water to flow into the well to replace the water being pumped out. This removes water from storage in the aquifer and produces a new gradient in the water table around the well, as shown in Figure 7.21. Viewed in three dimensions, this new configuration of the water table takes the shape of a cone and is called the *cone of depression.* The outer boundary of this cone will continually move into the aquifer until the amount of water being pumped balances water flowing through the aquifer toward the well. Theoretically the cone of depression will continue to move outward until it intercepts an area of natural recharge or discharge. In a thin aquifer of limited extent this may happen shortly, perhaps in a few months or years. In a thick,

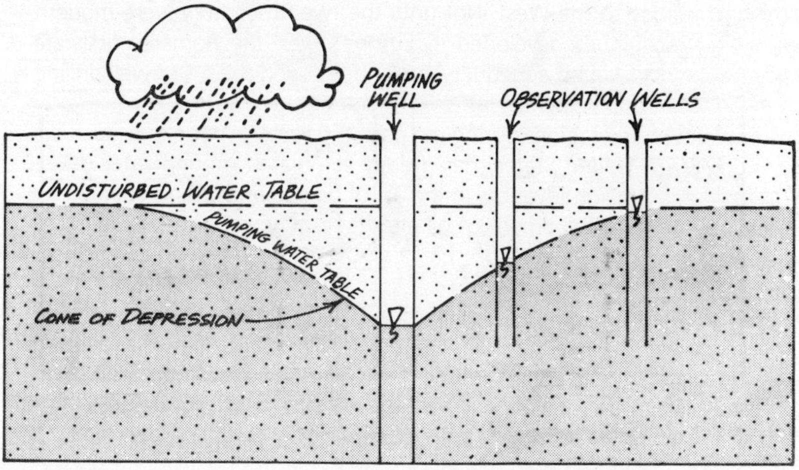

FIGURE 7.21 The cone of depression around a pumping well.

areally extensive aquifer, the cone of depression may continue moving out for a very long time, perhaps for decades or even centuries.

Observing the shape and extent of the cone of depression can teach a groundwater hydrologist a great deal about an aquifer's water-yielding properties. Observation wells located around a pumping well provide information on the shape of the cone by indicating the drawdown at each location. *Drawdown* is the depth that pumping has lowered the water table in the aquifer. Substituting data on drawdown, pumping rate, and time since pumping began into a well-known mathematical equation can provide the aquifer's approximate permeability and storage coefficient in the region around the well. This information is crucial in the design of future production wells. Once the equation values are known, they can help predict future drawdown in the aquifer at any distance from the pumping well. This is valuable information for planning to develop a water supply using an arrangement of several wells in a well field. Each aquifer is unique and will require optimal spacing among wells to produce the maximum water volume most efficiently.

The mathematical analysis of the cone of depression is beyond the scope of this book, but Figure 7.22 illustrates its important role. The three pumping wells are all producing at the same rate, and the shapes of the three cones of depression reveal how the aquifers respond to this pumping rate. The well in coarse sand and gravel clearly could be pumped at a much higher rate, and the well in fine sand could be pumped a little harder, but the well

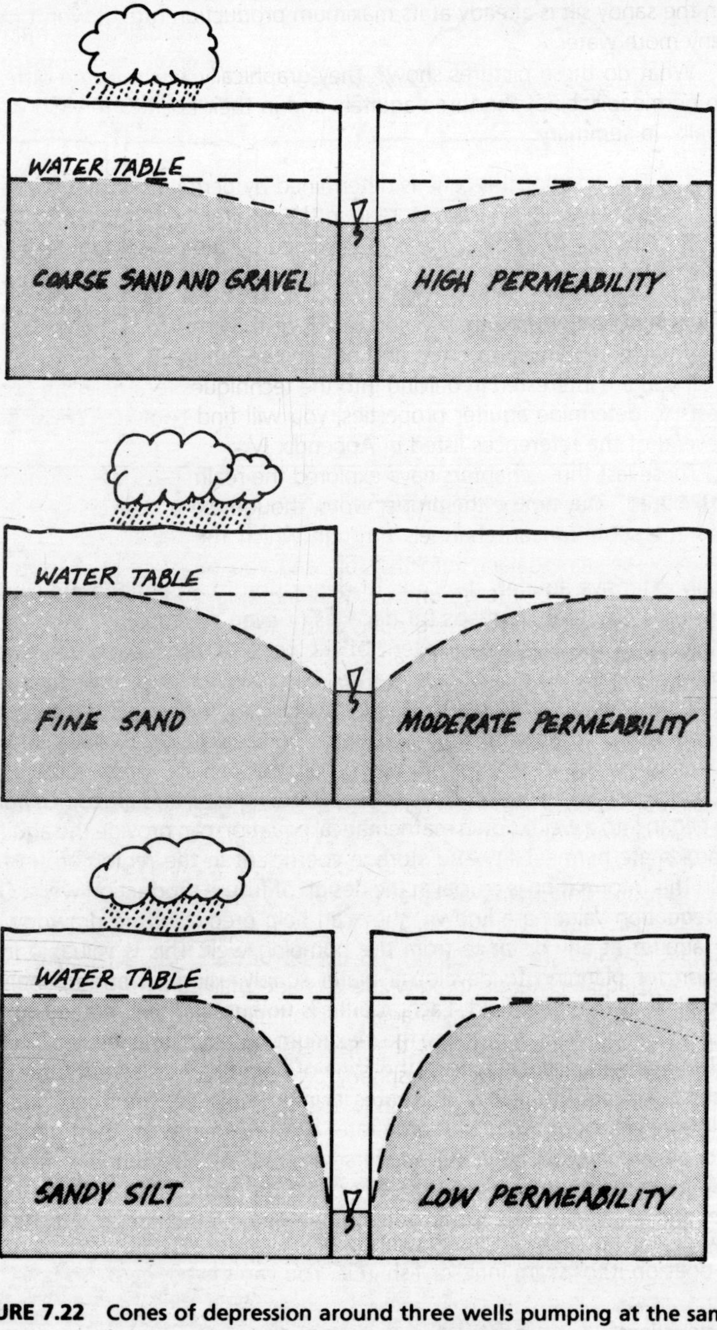

FIGURE 7.22 Cones of depression around three wells pumping at the same constant rate from aquifers with different permeabilities.

in the sandy silt is already at its maximum production rate. It won't produce any more water.

What do these pictures show? They graphically illustrate the differences in permeabilities of the three aquifers and in their abilities to yield water to wells. In summary:

1. The *slope* of the cone is determined by permeability and quantity of water removed from storage in the aquifer.
2. The *depth* of the cone is determined by the pumping rate.
3. The *radius* of the cone is determined by time since pumping began.

Theoretically, the growth rate and radius of the cone of depression depend only on the aquifer properties and not on the pumping rate.

If you are interested in delving into the techniques for analyzing pumping tests to determine aquifer properties, you will find complete discussions in several of the references listed in Appendix IV.

These last three chapters have explored the realm of underground water. "Explored" may not be the proper word, though, because you never actually see the subterranean channels through which the water flows. However, you do see rain soaking into the soil, and you see springs of clear water bubbling up out of the ground and flowing away in the creeks and rivers. That the rain going into the ground and spring water coming out are connected seems to us now to be an inescapable conclusion. But that was not always so. In fact, the discovery of this relationship between infiltration and spring flow (along with studies of evaporation) firmly established the concept of the water cycle just a little more than 300 years ago. Since that time we have learned about how water behaves underground, although no one has been able to observe it directly. The knowledge has come by inference through careful observation and records of rainfall, groundwater levels, and stream flow.

From our standpoint, a major function of the underground reservoir (in the overall water cycle) is to supply water to growing vegetation and to surface streams. To accomplish this the porous and permeable material of soils and aquifers tends to slow the passage of water through the underground reservoir, allowing storage to accumulate and keeping a backlog of supply always moving toward the discharge points. Its effect is to convert an intermittent supply of atmospheric precipitation to a continuous supply of soil water for plants and groundwater for streams.

APPLICATIONS

What kinds of applications can you think of for groundwater? You can't sail a boat on it, or swim in it, or fish in it. You can't even see it. So how could you use it? If you live in an area with no reliable surface-water supply, the answer is obvious. The best application for groundwater would be to supply all your water needs. It would be hard to think of an application more practical than that. In most places, in the United States at least, you would be able to supply your needs simply by drilling a well and installing a pump. The U.S. Geological Survey has estimated that the total usable supply of groundwater in the United States represents about 10 years' annual precipitation, or 35 years' runoff. Groundwater now accounts for about 20% of the total water used in the country; more than 11 million private wells supplement those used for public supply.

Groundwater has always been important to people living on farms or ranches beyond the reach of a public water system. During settlement of the West, one of the first things homesteaders did was dig a well to supply the household water needs. Where the water table was shallow, a hand-dug well often sufficed for domestic use; it required using either a bucket or a hand-operated pump. For deeper water tables, often more than a hand-

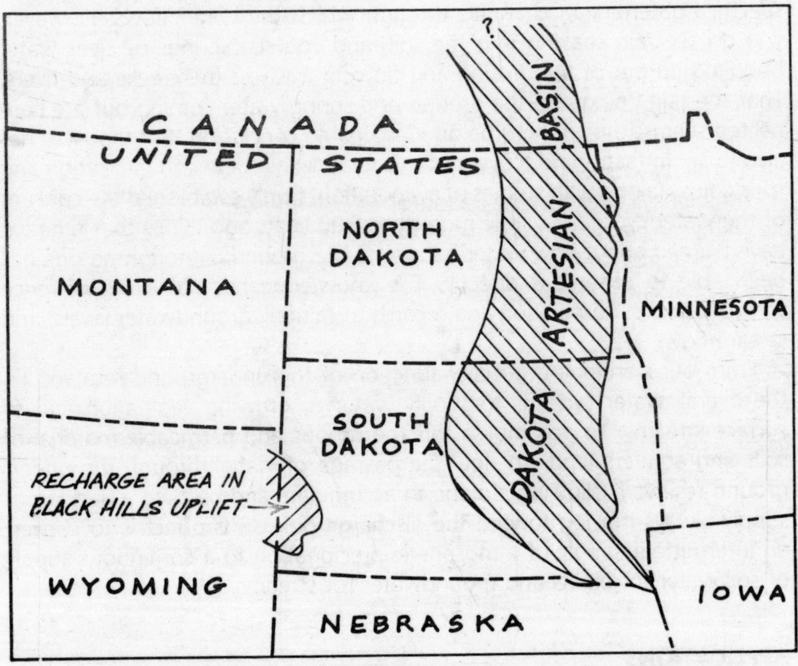

FIGURE 7.23 **Map showing Dakota Artesian Basin. Aquifers are recharged where they are exposed around the Black Hills uplift (adapted from maps by the U.S. Geological Survey and the North Dakota Geological Survey).**

operated pump was required to raise the water. Invention of the windmill pump in the 1850s solved this problem, and until the advent of small gasoline engines in the early days of this century, the steel windmill tower was a common sight on western farms, especially on the Great Plains. The northern Great Plains had something even better than windmills: There, artesian aquifers delivered water to the surface without mechanical power.

The Dakota Artesian Basin

The first artesian well was drilled in South Dakota in 1881, and by the end of the century hundreds of flowing wells were producing from aquifers in the Dakota Artesian Basin (see Figure 7.23). In the century since the first well was drilled, about 15,000 artesian wells have been drilled in South Dakota alone, thousands more in the North Dakota portion of the artesian basin. Many of these wells have stopped flowing and many have been abandoned. Some of the abandoned wells are still flowing, however, wasting water and gradually dissipating the gravitational potential energy stored in the aquifer. This potential energy results from the high-elevation recharge areas along the Black Hills uplift and the confinement of water in the downdip parts of the aquifers (see Figure 7.24). In its virgin state, the artesian system contained water under high pressure, which, when controlled at the surface, was capable of doing work. Several early-day flour mills were powered by water turbines operated solely by flowing artesian wells. Many small electric generating plants were also operated by flowing wells.

The chemical quality of the Dakota artesian water is not the best. Although the water can be used satisfactorily for irrigating crops, it is too highly mineralized to be considered good drinking water, according to U.S. Public Health Service standards. People, however, are sometimes surprisingly adaptable. Visitors from outside may turn up their noses at what they consider distaste-

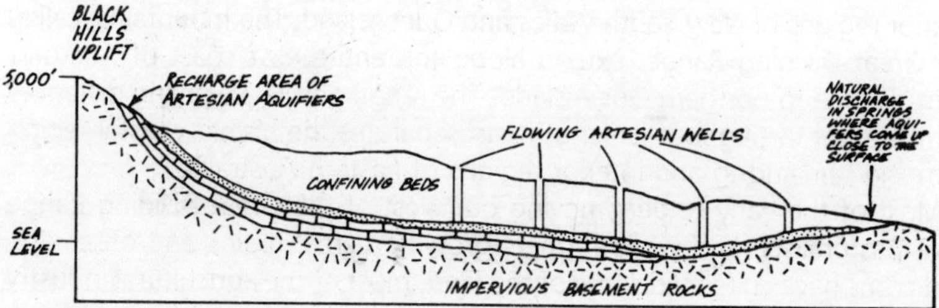

FIGURE 7.24 **Generalized east-west cross section of Dakota Artesian Basin (Adapted from publications of the U.S. Geological Survey).**

ful, unacceptable water, but when it is the only supply, residents either get used to it or move. The early settlers came to stay, and they weren't about to leave their land because the water didn't taste good. The generations of healthy, vigorous people who have grown up on this water during the last century prove that, as long as it contains nothing toxic, "highly mineralized" water may be beneficial for most forms of life, including children.

Without abundant groundwater, the Great Plains would have been settled much more slowly, and large areas probably would remain empty even today. These same conditions have also influenced settlement in other lands. An interesting parallel is found half a world away from Dakota, "down under" in the state of Queensland, Australia. An artesian basin there saved the day for early settlers. This one, called The Great Artesian Basin, had a history of development not unlike the Dakota Artesian Basin, but it was much more extensive and contained a much larger groundwater reserve.

The Great Artesian Basin

As the map in Figure 7.25 shows, this basin is aptly named. It is indeed "great" in every sense of the word. The basin underlies an area of 1,700,000 square kilometers (656,420 mi²), about one-fifth the area of the entire continent of Australia. As the area of the conterminous United States is almost the same as that of Australia, the Great Artesian Basin is also about one-fifth the area of the United States (excluding Alaska). To put it in even clearer perspective, the area of the Great Artesian Basin is about $2\frac{1}{2}$ times the area of Texas and is even about 10% larger than the entire land area of Alaska. The Dakota Artesian Basin would fit into a small corner of this enormous Australian basin.

The first settlement of Australia by Europeans came near the end of the eighteenth century with the establishment of English prison colonies along the southeast coast. Although exploration of the interior of the continent began early in the nineteenth century, it was several decades before permanent settlers pushed westward beyond the coastal mountains toward the interior regions of New South Wales and Queensland. The mountains, called the Great Dividing Range, extend along the entire east coast of Australia from Victoria to northern Queensland. The range occurs at varying distances from the sea throughout its length and separates the humid coastal region from the semiarid-to-arid interior regions of eastern Australia.

Most of the early settlers moving out west of the Great Dividing Range were pastoralists, and while farmers growing cereal grains and other dryland crops have cultivated some of the land, most of the agricultural industry is still devoted to livestock. Vast regions, particularly in western Queensland, contain no perennial streams, and the only reliable water source is groundwater from springs or wells.

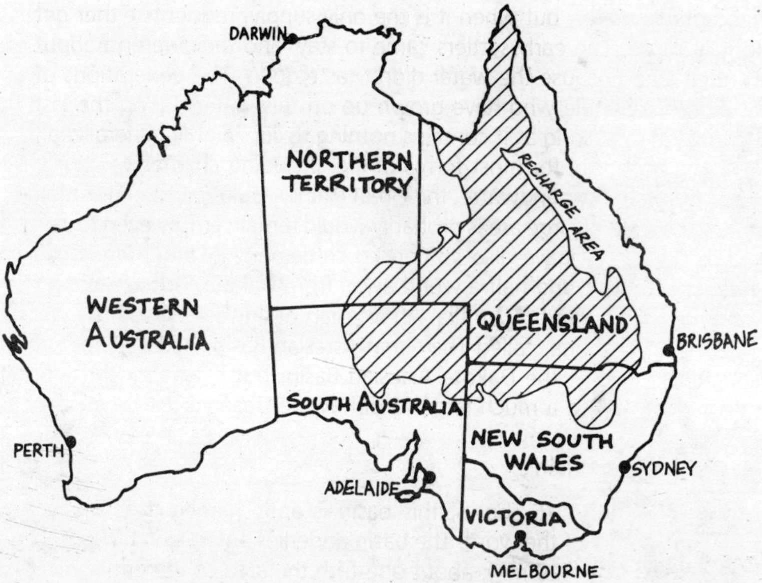

FIGURE 7.25 Map of the Great Artesian Basin, including recharge area at margin of basin along western slopes of the Great Dividing Range in eastern Queensland (source: Australian Bureau of Mineral Resources).

This is also a land of uncertain rainfall, making drought a constant worry. During the past century a major drought has occurred somewhere in Australia every 10 or 12 years, some lasting for almost a decade. In the 1895–1903 drought, the most severe in Australian history, the sheep population shrank about 50% and cattle more than 40%. Without wells to water their stock, the cattle and sheep ranchers in much of western Queensland would have been wiped out completely.

Although the land and mode of settlement in Dakota and Queensland are very different, the timing of the settlement and development of underground water coincided on both continents. The first artesian well was drilled in Queensland around 1880 and in Dakota in 1881. About 4700 flowing artesian wells have been drilled in the Great Artesian Basin during the past century. A typical early-day artesian well is shown in Figure 7.26. The well is on Yarrawin station (Americans would say "ranch") in northern New South Wales.

Because of extreme depths in the main artesian aquifers out in the middle of the basin, most of the flowing wells cluster around the margins. Nevertheless, the wells are generally deeper than those in Dakota. Although some

FIGURE 7.26 Artesian well at Yarrawin Station, New South Wales. Well is flowing at a rate of 478 gallons per minute (30 L/s) from a total depth of 1472 feet (449 m) (photo by E. F. Pittman (1907); Geological Survey of New South Wales).

flowing Dakota wells are over 2000 feet (610 m) deep, most average around 1000–1200 feet (300–360 m) or less. Flowing wells in Queensland, on the other hand, average about 500 meters (1640 ft) and some are as much as 2000 meters (6562 ft) deep. Figure 7.27 is a generalized cross section of the basin showing relative positions of the main aquifers and confining beds. Natural flow through the basin is from recharge areas along the west slope of the Great Dividing Range toward groups of springs around the margins of the basin where the aquifers again come to the surface.

The chemical quality of the water is much better than that in Dakota, and the water is widely used for domestic purposes as well as for watering livestock. Many small towns in the outback of Queensland and New South Wales depend entirely on artesian wells for their water. Unfortunately, the concentration of sodium in the water, while not high in absolute terms, is too high relative to the other dissolved constituents to make this water satisfactory for irrigation. As discussed in Chapter 5, when the relative percentage of sodium ions in irrigation water exceeds a certain value, the water can deflocculate the clay colloids in the soil and prevent infiltration of the

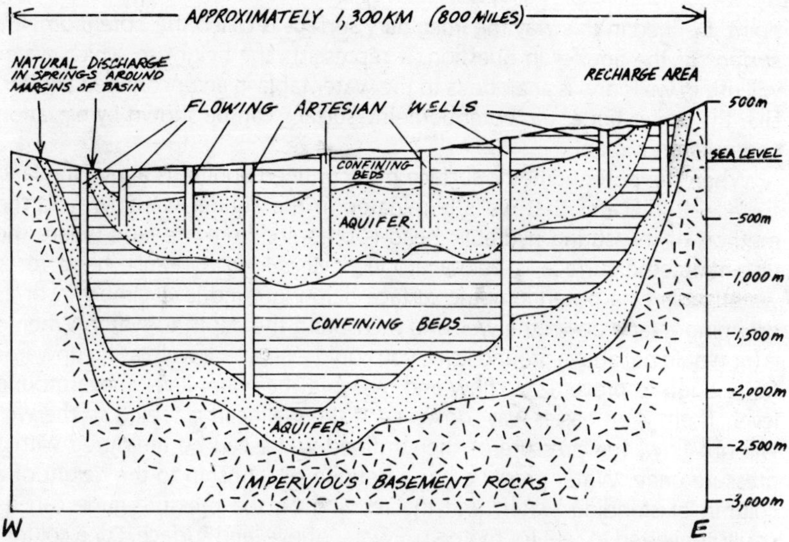

FIGURE 7.27 Generalized east-west cross section of Great Artesian Basin (adapted from publications of Australian Bureau of Mineral Resources).

soil surface. This is true of water in the Great Artesian Basin, and it partially explains the predominance of livestock raising over cultivated crops throughout the region served by the artesian aquifers.

Why Do Artesian Wells Flow?

As explained in Figure 7.16, the position of the hydraulic head in a confined aquifer relative to the land surface determines whether a well will flow. When the head drops below the land surface, the well is often referred to as a nonflowing artesian well. The Great Artesian Basin contains about 20,000 nonflowing artesian wells, most of which are equipped with windmill-operated pumps. The Dakota basin also has many thousands of nonflowing artesian wells.

When an artesian basin is first developed by wells, the hydraulic head is at a maximum. In its virgin state the basin has a level of hydraulic head progressively lower from areas of recharge in the hills toward areas of natural discharge (springs) in the distant lowlands, where the aquifers again come near the surface. *Hydraulic head* (the standing level of water in the wells) is a measure of the energy pushing water through the aquifers from recharge areas toward discharge points. You could think of this level of hydraulic head as an imaginary surface extending across the basin and expressing at any point the state of gravitational potential energy in the aquifer beneath that point. Defined in this way the imaginary surface is called the *potentiometric surface* of the aquifer in question. It represents the height to which water will rise in wells and is analogous to the water table in an unconfined aquifer. Like the water table, the potentiometric surface can be shown by elevation contours on a map.

When the potentiometric surface of a confined aquifer is above the land surface, it indicates that wells terminated in the aquifer will flow, and the method of measuring hydraulic head is different from the case where the potentiometric surface is below ground. Measuring hydraulic head for a water table or a potentiometric surface below ground level simply involves dropping a tape down the well and measuring the depth to water. When it is (or would be) above ground level, you could extend the well casing upwards far enough to prevent overflow and measure the water level above ground level. There is an easier way, however. If the wellhead is sealed so the well will not flow, the pressure of the confined water can be measured with a pressure gage. Water pressure has a definite relationship to the height of a column of standing water, and the reading on the pressure gage can be easily converted to feet (or meters) of water above land surface. For a column of water (as in a well casing), each foot of height produces a pressure of 0.43 pounds per square inch (2.99 kPa) at the base of the column. Each meter of water height produces a pressure of 9.81 kPa (1.42 lb/in.^{2}). For example, when artesian wells were first drilled in Dakota, some wells had a shut-in pressure of as much as 175 pounds per square inch (1207 kPa), equal to that of a column of water about 400 feet (122 m) above land surface. This was an extreme value, but it shows why pressure gages are used to measure potentiometric levels of artesian aquifers.

As an artesian basin is developed and more and more wells are allowed to flow, both the gravitational potential energy of the aquifer and the potentiometric surface are lowered. This has occurred both in the Dakota Basin and the Great Artesian Basin during the past century. Many wells were allowed to flow continuously, wasting water and gradually dissipating the aquifers' potential energy. As a result, some wells have stopped flowing, and the yields of those still flowing have been reduced. In eastern North Dakota the average yield from flowing wells is now only a few gallons per minute. Although many well yields are correspondingly low in the Great Artesian Basin, a number of flowing wells still have yields of 500 gallons per minute (32 L/s) or more. Australian hydrologists believe that conditions of head and flow are now at an equilibrium state in the Great Artesian Basin. Unless many more flowing wells are drilled, the potentiometric surface, which is above ground level in most of the basin, probably is now stabilized at its present level. Recharge and discharge appear to be in balance for the basin as a whole.

Both the Dakota Artesian Basin and the Great Artesian Basin are examples of large basins where the application of knowledge about groundwater involves simply drilling wells to tap a natural storage reservoir. Nature puts the water in and we withdraw it for our own uses. The volumes of water in storage are so enormous that the relatively small withdrawals, while they may lower the potentiometric surface, have little effect on overall storage. In small basins, on the other hand, heavy withdrawals for a prolonged period may seriously deplete the storage. Replenishing storage in these small basins is another way that we can apply groundwater knowledge to our benefit.

Storing Surplus Water Underground

As stated above, when someone mentions practical applications for groundwater, your first thought is apt to be of a well discharging water into a ditch or pipeline. This would be an accurate picture of an application in Queensland, or Dakota, or a thousand other places where the underground reservoir is being drawn on for a water supply. What about an application in which water is added to the reservoir rather than taken out? Nature has been doing that all along—natural groundwater recharge is an integral part of the water cycle.

We have learned from nature, and in doing so we have discovered some important advantages in using this subterranean reservoir. First, we know that no water is lost from storage by evaporation. Most of the time subsurface leakage out of the reservoir causes minimal loss, but if there is leakage, the flow rate underground is usually so slow that most of the stored water can be salvaged for use anyway. Except for the area devoted to infiltration ponds, no real estate must be purchased for the reservoir, and there are no construction costs except for dikes around the infiltration ponds and for wells to bring water back from storage. Finally it is often possible, as at Arvin-Edison, to store water in a reservoir beneath the land where the water will eventually be used. This saves conveyance costs (except for pumping) and saves the water almost inevitably lost during conveyance from reservoir to point of use. You may remember the Twin Lake Reservoir in Colorado cited in Chapter 3, in which conveyance losses were estimated at 16%–28% of the water released from the reservoir. A description of the infiltration ponds at Arvin-Edison demonstrates the practicality of this application.

Arvin-Edison Water Storage District. As described in the Applications section at the end of Chapter 3, the Arvin-Edison district is located along the southeastern margin of California's San Joaquin Valley (see map in Appendix I). Of a total 132,000 acres (53,420 ha) of land within the district, about 112,000 acres (45,326 ha) are devoted to irrigated crops. Historically all irrigation water was drawn from the groundwater reservoir beneath the district. With annual withdrawals far in excess of natural recharge, groundwater levels declined about 8 feet (2.4 m) per year until pumping lifts finally exceeded 600 feet (183 m) in many of the wells. The overdraft on the reservoir was estimated to be as much as 200,000 acre-feet (246,700,000 m^3) per year by 1960. It was obvious that sustaining farming operations would require an imported water supply, and in 1962 a contract was signed with the U.S. Bureau of Reclamation that provided for an annual supply of surface water to be brought by canal from a reservoir on the San Joaquin River east of Fresno.

As described in Chapter 3, the supply consists of Class 1 (firm) and Class 2 (nonfirm) water. The Bureau of Reclamation has promised to deliver at least as much water each year as the Class 1 entitlement calls for. Class 2 water, on the other hand, is water in excess of all Class 1 commitments that may be available from the surface reservoir on a seasonal basis. The Arvin-Edison allotment of Class 1 water is 40,000 acre-feet (49,340,000 m^3) per year and Class 2 water, when available, is provided up to 313,000 acre-feet (386,085,500 m^3) per year. The two classes of water are predicated on normal precipitation and runoff conditions. In years when drought conditions prevail, even the Class 1 entitlement may be reduced for all contractors. For example, in the drought year of 1977, the bureau allotted Arvin-Edison only about 10,000 acre-feet (12,335,000 m^3) of Class 1 water and no Class 2 water. In wet years, when the river reservoir is full and the snowpack promises a large spring runoff, substantial deliveries of Class 2 water are made during the winter months. Over the long term it is estimated that the average annual supply of Class 1 and Class 2 water will be about 191,000 acre-feet (235,598,500 m^3).

When farmers need water during the irrigation season, contract water being delivered to the district is distributed directly to the land. Water in excess of immediate irrigation needs is put into the infiltration ponds for eventual storage underground. Because of the extreme variability of precipitation and its associated runoff in this part of California, the annual surface-water supply is erratic and uncertain. For examples of this variability, examine the long-term records of precipitation at Bakersfield (Figure 4.13) and flow in the nearby Kern River (Figure 8.21). The San Joaquin River (with a reservoir east of Fresno) that serves Arvin-Edison has a similar pattern of variation. To overcome the hazards of depending on an erratic supply of surface water and a seriously depleted groundwater reservoir, the Arvin-Edison project was designed to use both surface water and groundwater to meet irrigation needs. Using underground storage to smooth out the extremes, the district has been able to provide a dependable supply of surface water to about 40% of the irrigated land. Concurrently the regulation of underground storage allows irrigation of the remaining lands with groundwater indefinitely. Figure 7.28 shows how the underground storage of excess surface water

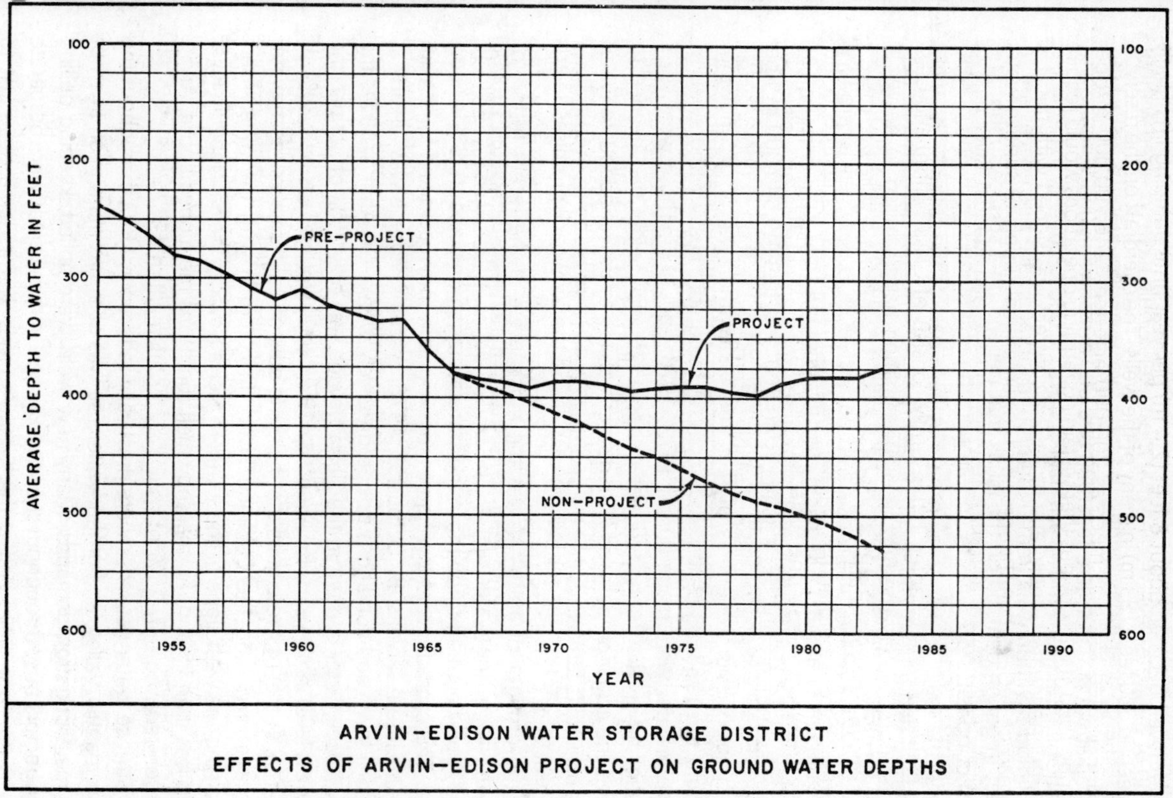

FIGURE 7.28 Effects of Arvin-Edison Project on groundwater depths (courtesy of Arvin-Edison Water Storage District).

has stabilized water levels and eliminated the overdraft on the groundwater supply. The dashed line on the graph labeled "Non-Project" is an extrapolation of the water-level decline that most likely would have occurred without the recharge operations. During its first 17 years of operation the district has stored about 800,000 acre-feet (986,800,000 m³) of water underground and has delivered almost 2 million acre-feet (2,467,000,000 m³) directly to farmers.

Periodic withdrawals from underground storage have been necessary to meet irrigation demand on the lands that depend on a surface supply. The district operates 50 wells in and around the infiltration ponds and 5 wells along the main supply canal in the northern part of the district. The wells range in depth from 750 to 1,078 feet (229–329 m) and have 16-inch (41-cm) diameter inner steel casings. Each well is equipped with a 300 horsepower (224 kW) electrically-powered pump, which is designed to produce a minimum of about 1800 gallons per minute (114 L/s).

Figure 7.29 is a groundwater hydrograph from an observation well about a quarter mile (0.4 km) north of the Sycamore infiltration ponds. This hydrograph shows the effects of recharge and withdrawal of groundwater during the 1966–83 period. As shown on the graph, heavy withdrawals of groundwater were made during 1972, 1976, and 1977. Those were years when the district's imported water supply was well below that needed to meet the irrigation demand. The district's ability to deliver adequate water supplies to its farmers during the severe drought of 1976–77 was particularly noteworthy.

The drought of 1976–77 was the most severe in the recent history of California. In the water year 1976–77 (October 1, 1976 to September 30, 1977) precipitation throughout the state averaged only about 35% of normal. This caused record low figures for mountain snowpacks and for stream runoff into reservoirs. By October 1, 1976, reservoir storage was already at record lows, and conditions got progressively worse. On April 1, 1977, the snowpack water content in the Sierra Nevada was only 25% of normal, the lowest level in 47 years. By August 1, 1977, total storage was only 39% of normal in the 143 reservoirs that held the bulk of California's surface-water storage (excluding the Colorado River). In common with the other rivers draining the west slope of the southern Sierra Nevada, the San Joaquin River had less than 25% of normal snowmelt runoff. As mentioned above, Arvin-Edison received only 25% of its Class 1 entitlement in 1977.

Surface water deliveries were drastically reduced in 1977 throughout the San Joaquin Valley, and groundwater use was at a maximum in an attempt to meet irrigation demands. Many new wells were drilled, keeping drilling contractors busy throughout the year. Farmers in many areas were desperate for water, not only to grow the season's crops, but in some cases just to prevent permanent damage to established orchards and vineyards. Ground-

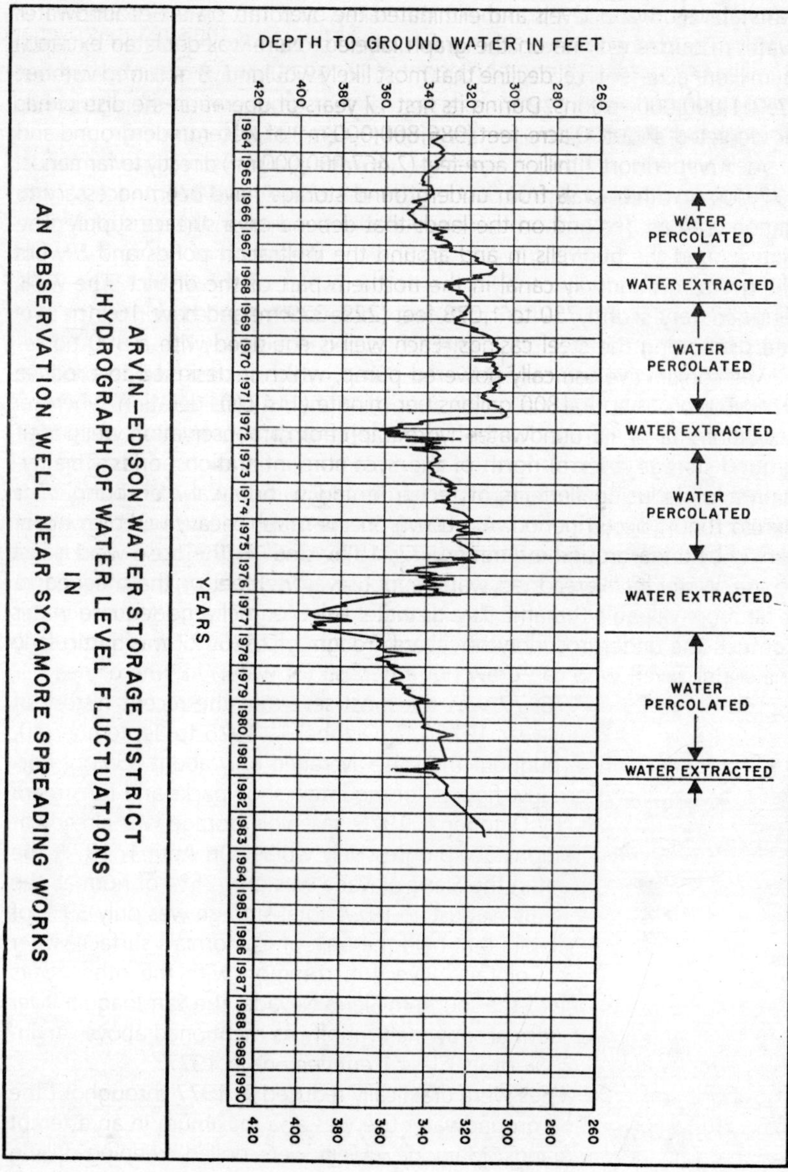

FIGURE 7.29 Hydrograph of water-level fluctuations in an observation well near Sycamore infiltration ponds (courtesy of Arvin-Edison Water Storage District).

water levels dropped throughout the valley, and the state department of water resources estimated that underground storage was depleted by about 3 million acre-feet (3,700,500,000 m³) in 1976 and 6 million acre-feet (7,401,000,000 m³) in 1977. In a normal year groundwater storage would be depleted about 1.2 million acre-feet (1,480,200,000 m³).

At Arvin-Edison the net reduction in underground storage was almost 127,000 acre-feet (156,654,500 m³) for the combined 1976 and 1977 irrigation seasons. The groundwater level declined sharply in the cone of depression around the pumping wells, as shown in the graph of Figure 7.29; but the overall groundwater level beneath the district, as shown in Figure 7.28, declined very slightly. When recharge operations resumed during the wet years following the drought, water levels began a steady rise.

The conjunctive use concept, exemplified by the successful operations at Arvin-Edison, is a practical application that merits consideration wherever conditions favor its use. Under favorable geological conditions an underground storage reservoir is often the most efficient and cost-effective alternative for reducing fluctuations in an otherwise erratic water supply. The excess runoff occurring in wet years is inexpensive, because if it isn't conserved by underground infiltration, it just flows out to the ocean and is lost to use. Water recovered from storage for use in dry years, on the other hand, is far more valuable than the surplus water used for recharge. Viewed in this context, the underground water "bank" actually pays a form of interest to the water saver, who can draw out a deposit (of water) in time of need.

8

Runoff

. . . even the weariest river winds somewhere safe to sea

(A. C. Swinburne)

Can you imagine a river ever being weary? Apparently the poet, Swinburne, could. As a matter of fact, poets and songwriters seem to find rivers good subjects for metaphor ("down eternity's river," "Alph, the sacred river," "a golden river," "Old Man River," etc.). And once a river acquires a metaphorical identity, the idea of a river getting weary isn't all that farfetched, either. Have you seen anything else in nature in constant motion, moving in the same direction, day and night, year in, year out? You can feel the wind, even if you can't see it; but the wind changes direction and is sometimes still. Waves on the sea are generated by wind, and while they break endlessly on the shore, they too change direction occasionally. A river, though it may wander on its floodplain, always flows in one direction—from the highlands to the lowlands and on to the sea. Only when great geological forces rearrange the earth's crust and change the tilt of the land are rivers ever likely to change their direction of flow.

Where the crust is stable, as in Australia, the history of rivers is measured in terms of geologic time. The Finke River, an intermittent stream that drains a large area in central Australia, has been following the same course from the highlands of the MacDonnell Ranges to a sink in the Simpson Desert for hundreds of thousands, perhaps even millions of years. Although the Finke is one of the world's oldest rivers, many others have histories extending back in time at least to the last ice age. This constancy of flow and location has made rivers important to people. All the great cities of antiquity were on the sea coast or on rivers, and those on the coast were often at the mouths of rivers.

The story of runoff as a phase in the water cycle is the story of water flowing across the land in streams and rivers. The total process includes several intermediate steps between precipitation and stream flow. While our visual image of runoff is the running water flowing between the banks of a stream, the intermediate steps are also important. To understand the runoff process fully requires looking at what happens between the arrival of raindrops on the land and final delivery of water into the streams that carry it away to the sea. This sequence of events is sometimes called the *runoff cycle*.

THE RUNOFF CYCLE

The runoff cycle can be explained by following a raindrop as it comes to the earth and travels toward the nearest stream. Unless this errant raindrop lands on the surface of the stream itself, it may travel a circuitous route before it joins the flowing water of the stream. Much of what happens to rain between the time it falls and its entry into a stream can be observed directly, but some aspects of the runoff cycle must be inferred from indirect evidence. What follows is a sequential account of some possible intervening steps between rainfall and stream flow.

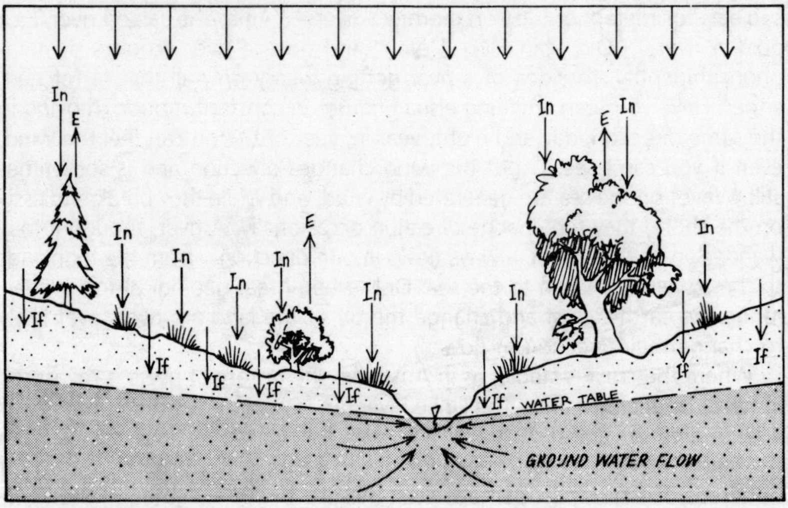

FIGURE 8.1 Early stage of runoff cycle (rain just beginning to fall).

This is an account of the runoff cycle in a humid region. The story begins after a spell of dry weather when the vegetation and soil surfaces are essentially dry. The stream is being fed by groundwater flowing into its banks and through its bed (see Figure 8.1). Rain from a new storm is just now beginning to fall. In the early part of the storm, much of the rain is intercepted (In) by vegetation, and the rain that falls on the ground is drawn into the soil by infiltration (If). Except for rain that falls on the stream surface or on bare rock surfaces, there is no immediate flow into the stream. Some of the intercepted water begins to evaporate (E) from the vegetation.

As rain continues to fall, it runs off the vegetation and falls to the ground. If it is coming down steadily, it may exceed the soil's infiltration capacity and begin to fill shallow depressions and to accumulate as a layer on the surface, which then begins to move as *overland flow* toward a watercourse, as shown in Figure 8.2. Water infiltrated into the soil will now begin to percolate downward toward the water table. The soil permeability may limit the amount that can percolate downward, however, and some water may begin to flow downhill just beneath the soil surface as a component of runoff called *interflow*. Water flowing as interflow moves in the same direction as overland flow on the surface, but in the subsurface it flows more slowly and arrives at the stream later than overland flow. Finally, as rain continues to fall, some water is still evaporating from the vegetation and the soil, but now the

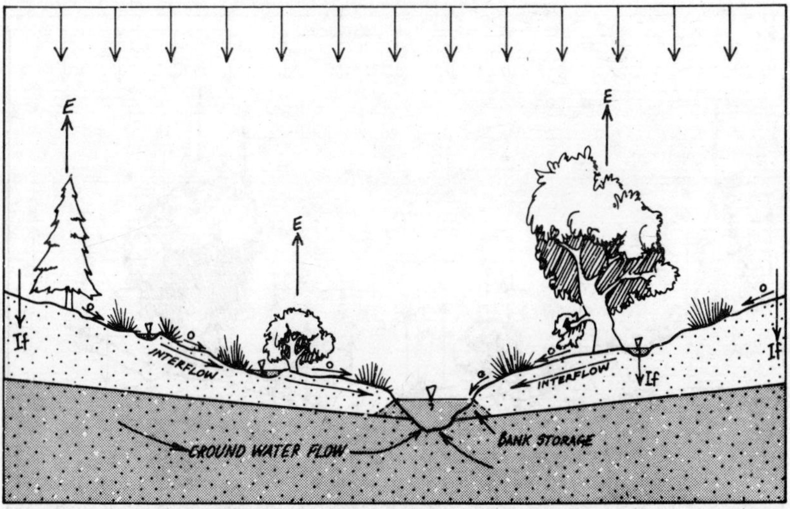

FIGURE 8.2 Second stage of runoff cycle (continuous rain is producing runoff from overland flow and interflow).

surface depressions are overflowing and water flows continuously toward the stream as overland flow and as interflow. If the groundwater is shallow, recharge from the rain may cause the water table to rise near the stream, causing additional groundwater to flow into the stream. In most cases, however, storm recharge to the groundwater enters the stream long after the storm is over, due to the slow rate of groundwater movement.

When the rain stops, overland flow and interflow soon cease, and evaporation removes water from the wet vegetation and surface puddles. Figure 8.3 shows the landscape beginning to dry. Note that the stream level is higher now than at the beginning of the storm. Water from the stream went into the banks as the stream rose and now will slowly run back out as the stream level drops toward its dry-weather level. This is called *bank storage.*

Bank storage is shown schematically in Figure 8.4. As the stream rises, water infiltrates the soil or rocks of the stream banks as a wedge-shaped body of water temporarily stored in the ground adjacent to the stream. After the storm, as the stream begins to fall, this water seeps out of the ground slowly and increases the stream flow. Where the geology is favorable and the stream banks are porous, a lot of storm runoff may be temporarily stored along the length of a stream. This allows retention of much potential floodwater in the upstream areas of a drainage basin and a corresponding reduction in downstream flood crests. Of course, this only works when the

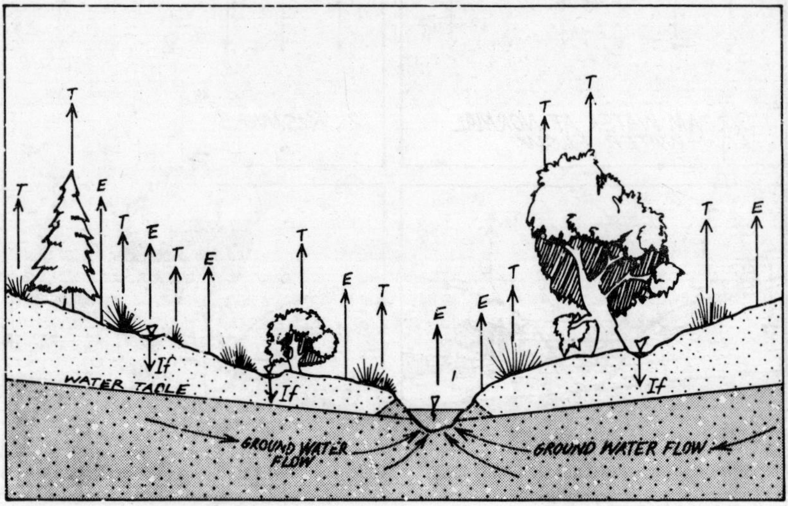

FIGURE 8.3 Runoff cycle ends as rain stops and landscape begins to dry out (transpiration, T; evaporation, E).

bank storage reservoir is empty. If one storm follows another in rapid succession, floods may be unavoidable.

The runoff cycle in an arid or semiarid region will differ from that of a humid region. In the former, the groundwater is usually deep in the ground, well below the stream beds. Much of the stream flow depends on precipitation alone, and since long, dry spells often separate storms, most streams are ephemeral, or at best intermittent. Years ago in New Mexico the country people used to say that you can tell where an arroyo (watercourse) is by the cloud of dust that hangs over it. Most of the time that's true, but during a summer thunderstorm the dusty arroyo may become a raging torrent. Another factor to consider in the runoff process, then, is the type of storm and its duration and intensity.

The three major types of storms described in Chapter 4—*cyclonic, orographic,* and *convective*—may occur in any climatic region; one or two usually

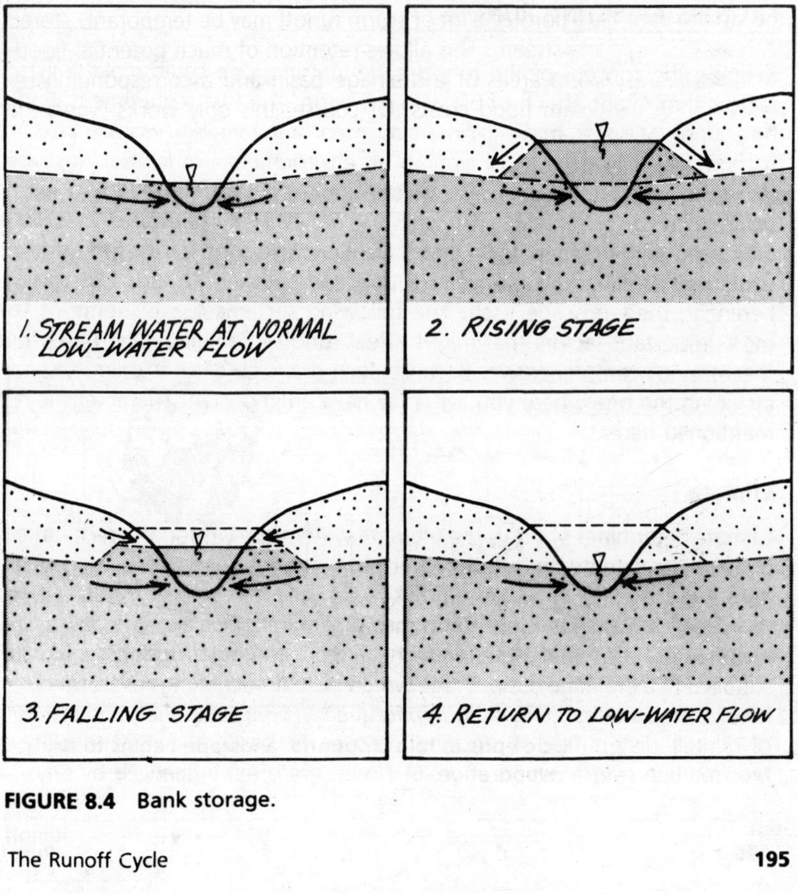

FIGURE 8.4 Bank storage.

predominate in a given area. Heavy rainfall may be produced in a cyclonic system or by orographic lifting, but storms of this type often bring only light to moderate rainfall. These are typical in humid regions. In arid or semiarid regions, on the other hand, most of the annual precipitation comes from convective storms, which often produce heavy rainfall for short periods. In the arid runoff cycle, surface flow begins shortly after rain begins to fall, and overland flow is the chief mechanism for draining water into the streams. The overland flow velocity often is higher than in a humid region, and water has little time to infiltrate the soil or stream banks. The desert also has little vegetation to intercept rainfall. Like the rainfall, the resultant runoff is intense and of short duration. Water doesn't linger in the stream channels, and even though most of these are losing streams, groundwater recharge from the streams is minimal. That, along with the scarcity of precipitation in general, is why water tables are usually deep beneath the desert floor and groundwater storage is meager, compared with humid regions.

FACTORS AFFECTING RUNOFF

Keeping the concept of the runoff cycle in mind, one could make a list of factors that might affect runoff within a basin. The first requirement would be a supply of water, making climate an important item. You wouldn't expect to find a river like the Mississippi in an environment like central Australia. However, even with an adequate water supply, runoff and streamflow might be influenced in a number of ways by the natural characteristics of the drainage basin. In the end you might want two lists, one for the atmospheric conditions affecting overall water supply and another listing the factors inherent in the landscape itself. The following sections discuss some of the more important factors that might affect runoff in any drainage basin, but they are not comprehensive. Because each drainage basin is unique, some, including the one where you live, may have an important characteristic not mentioned here.

Climate

Climate determines not only the ultimate water supply through precipitation, but also the extent to which that precipitation is returned to the atmosphere before it can participate in stream flow. As used here, climate is considered to include not only the long-term moisture supply, but also the day-to-day weather patterns that determine the timing and total quantity of water supplied to a drainage basin. These would include such things as storm types, kind of precipitation (rain or snow), frequency, typical duration and intensity of rainfall, distribution of precipitation over the basin, and dependability of precipitation over a period of years. Floods are often influenced by physical conditions within the drainage basin, but droughts are determined entirely by climate.

Where a lot of precipitation occurs, obviously a lot of runoff is also apt to occur, but the effect of timing is not always so obvious. As explained above, a storm following a period of dry weather causes a certain sequence of events that may produce runoff, depending on the physical characteristics of the basin. A series of storms in rapid succession results in a different set of conditions in which not only runoff, but also flooding, is very likely to occur. When the soil is full of water and storage has reached capacity in stream banks, further precipitation tends to cause immediate rises in stream levels. An important factor for predicting floods, then, is *antecedent precipitation,* which is an index of the moisture stored within a drainage basin at the beginning of a new storm.

Where much of the annual precipitation takes the form of snow, temperature plays a part in determining runoff patterns within a drainage basin. Snow falling on unfrozen ground will form an insulating blanket and prevent freezing of the soil. Then when melting begins in the spring, some of the resulting water will infiltrate the soil surface, slowing the runoff process. On the other hand, snow on frozen ground also insulates the surface and prevents early thawing of the soil. Frozen soil is relatively impervious, and most of the snowmelt will tend to run off by overland flow toward streams, causing stream levels to begin rising as soon as the snowpack begins to thaw. If you have a heavy snowpack on frozen ground up in the mountains, you hope for cool spring weather!

Physical Characteristics of the Drainage Basin

Terrain conditions affecting runoff can be considered in two categories: (1) conditions inherent in the natural landscape, and (2) conditions in which nature has been altered by human use of the land. Some of the natural features might be basin elevation and orientation, topography (shape and slope of the land), vegetation type and amount, soil type, and geology. The land-use practices having the most general effects on natural drainage are those related to agricultural or urban developments.

Elevation and Orientation of Basin. The main effects of elevation are related to temperature. At higher elevations, cooler temperatures result in less water loss by evapotranspiration. Above timberline there usually isn't much vegetation to transpire any water, and, of course, most of the annual precipitation comes as snowfall. In addition to the snowpack, water is also stored throughout the winter in frozen soil and in ice on lakes and streams. Most of the runoff from high basins comes between the spring thaw in April or May and the next freeze in October or November. The light winter runoff comes chiefly from groundwater seeping into streams below the frost line. The main variations on this general runoff regime in the high country are the result of basin orientation. In north-facing basins (in the Northern Hemisphere), snow often lasts well into summer, while in many south-facing basins snow melts early in the spring. Also, in some years, depending on time between storms, snow melts soon after it falls, leaving little for spring runoff in the southerly basins.

Basin orientation in relation to the prevailing storm tracks may also have a pronounced effect on runoff, especially on the timing of streamflow through the drainage net. As a storm center moves across the basin, precipitation begins and ends at different times in various parts of the basin. The direction of storm movement in relation to the direction of streamflow in the drainage net will determine when and where the highest stream flows occur, and also the total duration of runoff as storm waters move through the basin (see Figure 8.5). By moving an imaginary storm center across this basin in several directions, you can see how the concentration of precipitation from the storms would produce different patterns of runoff in the tributaries and in the main stream. From the standpoint of the water cycle, it doesn't matter. Runoff from precipitation always gets delivered somehow to the streams draining down to the rivers and the sea. To the hydrologist concerned with water supply and flood control, however, how the system responds to storm runoff matters a great deal. Besides a knowledge of the water cycle, hydrologists need an intimate knowledge of the characteristics of the drainage basins for which they are responsible.

Topography. The general shape of the landscape, the steepness of the slopes, and the total relief all affect the way precipitation reaches the streams in the drainage basin. As you would expect, the steeper the slopes, the quicker the tributaries feed storm runoff into the main stream draining the basin. An important aspect of topography is the areal configuration of the drainage basin itself; the general outline is shown on the map in Figure 8.5.

As stated previously, nearly all erosion is the result of running water, and present stream channels were formed by runoff from the land during past storms. The whole system is delicately balanced and responds quickly to environmental changes. This is particularly evident in many drainage basins in the western United States in which geologic (tectonic) forces are still active and the land is rising. As the land tilts up, the streams have more energy and actively erode their channels deeper into the underlying rocks. Eventually the stream channels will arrive at new gradients that will be in equilibrium with the water supply and the regional tilt of the land. Meanwhile, the drainage basin will continue to maintain its function in the water cycle of collecting and delivering runoff to the rivers flowing to the sea.

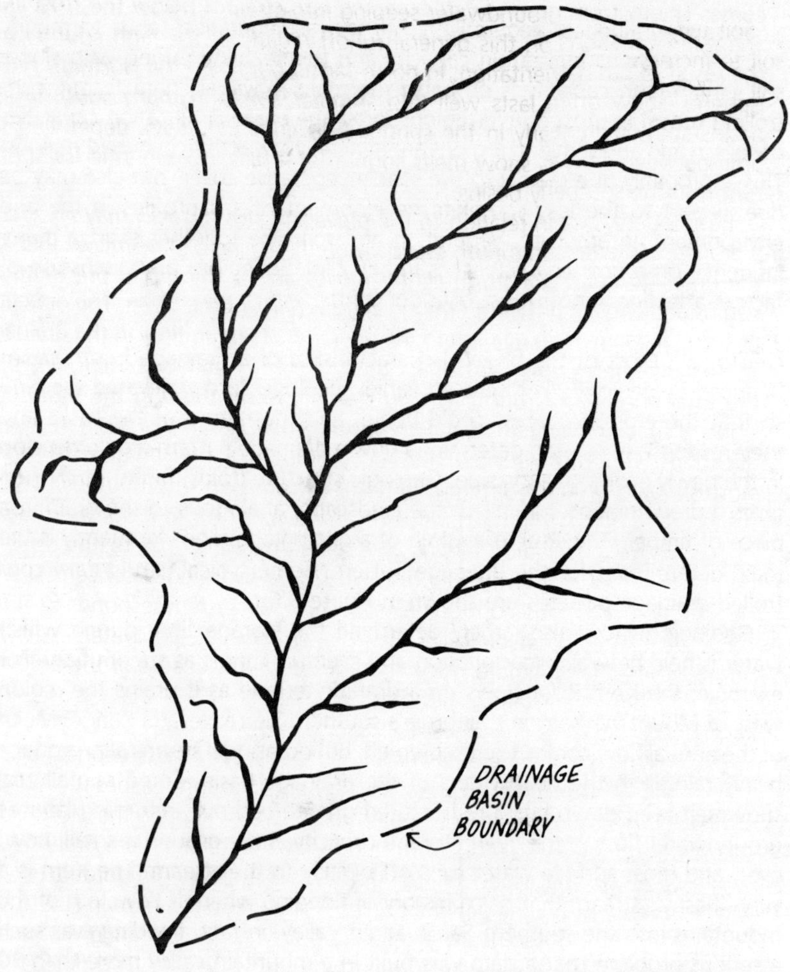

FIGURE 8.5 A typical drainage basin with a main stream and tributaries.

Vegetation and Soil Type. Vegetation is the chief agent of water loss (through transpiration) in the equation relating precipitation and runoff. The type and density of vegetation also influences how much precipitation is intercepted above the ground and how much infiltrates the soil once the water falls to earth. The existence of a vegetative cover retards overland flow, giving the water more time to infiltrate the soil. This can make a substantial difference in the rate at which stream levels rise during a storm.

Soil type influences vegetative cover just as vegetation tends to modify soil to increase its infiltration capacity and permeability. Along with slope, soil's infiltration capacity is probably the major physical characteristic controlling runoff in most humid regions' drainage basins. Infiltration capacities of soils in arid and semiarid regions are generally lower than in humid regions. This is probably due mainly to the sparse vegetative cover, but also may be due in part to the less complete development of soil profiles in the arid environment. In any case, vegetation and soil type together exert a major influence on runoff characteristics in any climatic region; any discussion of factors affecting runoff must consider both.

Geology. Most of the physical characteristics of a drainage basin are influenced by geology. As discussed earlier, geologic forces elevated the land so that the erosion process could begin, and the nature of the rocks and their response to erosion determined how a drainage pattern would develop in the newly evolving landscape. Geologic structure (folds, faults, etc.) often gives a predominant "grain" to the landscape, analogous to the grain in a piece of timber. The structural grain of a geologic terrane[1] frequently is the main determinant for the drainage pattern. Some typical, structurally controlled drainage patterns are shown in Figure 8.6.

Geologic factors also largely determine the storage time during which water is held between precipitation and eventual runoff as stream flow. For example, the Kern River flows on a granitic terrane as it drains the region west of Mount Whitney in California's southern Sierra Nevada Range. Much of the annual flow comes from snowmelt, but occasional storms also produce heavy rainfall in the headwaters of the drainage basin. Some rainfall and snowmelt seep down into the fractured granite bedrock, returning later as groundwater flow in the river. Storage capacity in the ground is small, however, and most surface water runs off quickly to the stream. The Kern is a very "flashy" stream that has a history of flooding where it flows out of the mountains into the southern San Joaquin Valley. In fact, flooding was such a serious problem that a dam was built in a mountain valley more than 30 years ago to control floodwaters and provide storage for irrigation water for valley farmers. The storage capacity behind the Isabella dam is about 540,000 acre-feet (666 million m^3), which in normal years is sufficient to store most of the spring snowmelt. The long-term average flow for the Kern is about 700,000 acre-feet (863 million m^3) per year. To show how unpredictable this stream is, consider an early December 1966 storm. In 48 hours more than 200,000 acre-feet (247 million m^3) entered the Isabella reservoir from a rainstorm on the upper Kern. Without the dam, a destructive flood would

[1]When used to describe landscape generally, the word is spelled *terrain*. A geologic *terrane* is a landscape underlaid by a predominant rock type (e.g., a granitic terrane, a volcanic terrane, a limestone terrane, etc.).

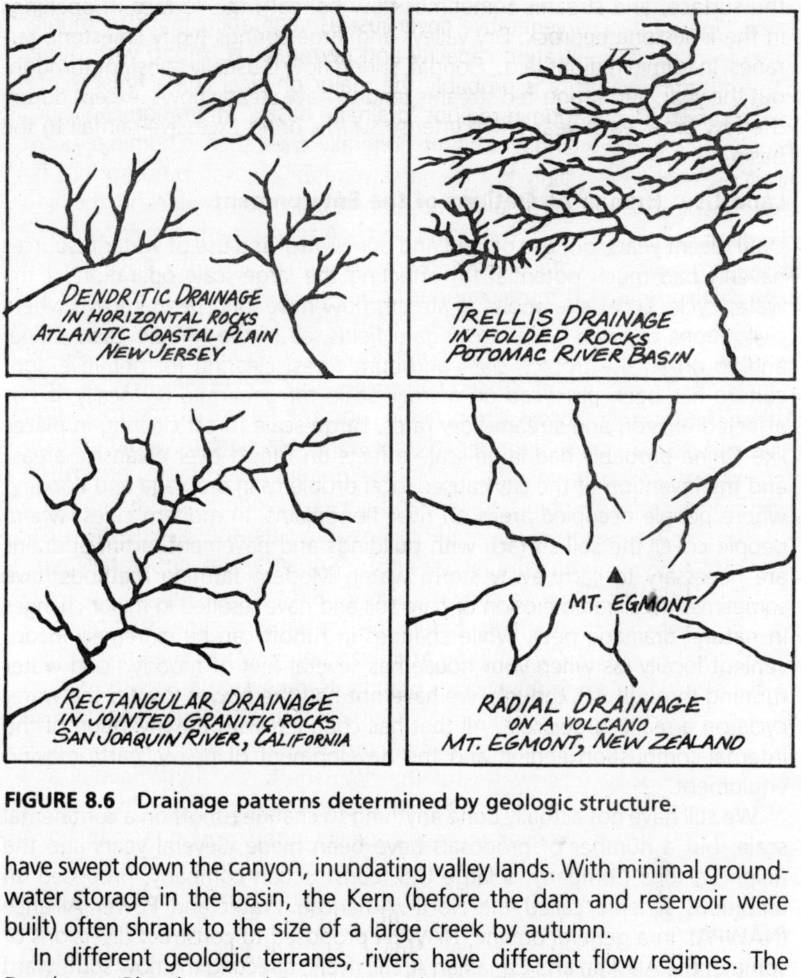

FIGURE 8.6 Drainage patterns determined by geologic structure.

have swept down the canyon, inundating valley lands. With minimal groundwater storage in the basin, the Kern (before the dam and reservoir were built) often shrank to the size of a large creek by autumn.

In different geologic terranes, rivers have different flow regimes. The Deschuttes River in central Oregon provides a marked contrast with the Kern. Flowing through a volcanic terrane, the Deschuttes has a very even flow throughout the year. In fact, it is considered to be one of the most stable rivers in the United States. Draining a region just east of the Cascade Range, the river obtains much of its flow from snowmelt, as does the Kern. Much of the land surface is underlain by thick deposits of volcanic ash and other very porous volcanic rocks. Most of the annual snowmelt or rainwater seeps into the ground and feeds the stream steadily throughout the year. Serious floods are extremely rare on the Deschuttes.

A limestone terrane is another kind of geology that provides characteristic stream-flow patterns. Limestone drainage basins contain many sinkholes at the surface, and streams sometimes flow beneath the surface in openings in the limestone bedrock. Dry valleys and large springs typify limestone terranes. In humid regions (e.g., Florida), spring flow is usually constant throughout the year, and spring-fed streams tend to have steady flows, except during the few times each year when intense storms bring excessive rainfall to the basin.

Land Use: Human Alterations of the Environment

Until recent years, our use of land and our control and use of water resources haven't had much potential for affecting the large-scale operation of the water cycle. Local alterations of stream flow have long been known where civilizations diverted water to irrigate fields, as in the Middle East, China, and so on. In the tropics, slash-and-burn forest clearing for primitive agriculture has been practiced on a small scale for generations; locally, it has affected erosion and stream-flow rates. Large-scale forest clearing in places like China probably had large-scale effects on runoff over extensive areas; and the invention of the city caused local problems in drainage and flooding where people occupied areas on river floodplains. In modern cities, where people cover the soil surface with buildings and pavement, artificial drains are necessary to carry away storm water. Modern farming methods have sometimes accelerated erosion of bare soil and have resulted in minor changes in natural drainage nets. While changes in runoff can be extremely inconvenient locally (as when your house has several feet of muddy flood water running through it!), though, we have not had the power to alter the water cycle on a really large scale. All that has changed with the invention of the internal-combustion engine and the development of heavy, earth-moving equipment.

We still have not actually done anything to change runoff on a continental scale, but a number of proposals have been made. Several years ago the Ralph Parsons Company, a large U.S. construction company, proposed an ambitious scheme called the North American Water and Power Alliance (NAWPA). In a general outline, NAWPA proposed to construct diversions of some of the Alaskan and Canadian Arctic rivers, directing the flow southward into southern Canada, the United States, and northern Mexico. This would deprive the Arctic Ocean of a large amount of runoff from North America. Concurrently, the Soviet Union proposed diverting the flow of some of its large, northerly flowing rivers into the dry lands of central Asia. Such diversion would also deprive the Arctic Ocean of much natural runoff. What effect these proposals would have on the global water cycle, no one can say. While neither is likely to happen, the point is that either could.

We may well have it within our power to alter the water cycle, and perhaps even the worldwide climate, radically. Much has been made of possible alteration of the climate through uncontrolled release of CO_2 from fuel combustion as well as damage to the earth's ozone layer through excessive release of gaseous fluorocarbons to the atmosphere. Releasing gases to the atmosphere is an unavoidable consequence of our technological civilization, but rearranging natural runoff on a continental scale is another matter entirely. Whether any of these things will ever come to pass is still an unresolved issue, though. Debate continues in the scientific community, and hard evidence of human-caused changes in climate still eludes us. The possibility offers a good reason, though, to be aware of the water cycle and all its ramifications and to realize the possibility of future change. It makes one more aware of some of the otherwise obscure news stories buried in the back pages of daily newspapers.

As a final word on factors affecting runoff, it should be emphasized that, while it is convenient to consider various factors individually, all of them operate collectively in a truly dynamic system. A slight change in one may magnify as it works its way through the system. A fire that destroys part of the ground cover, for example, may lead to faster overland flow, less infiltration, more erosion, less groundwater recharge, flooding in the stream, and so forth. Human records are not extensive enough to establish the existence of long-term climate cycles, but precipitation and stream-flow records do show sequences of abnormally wet or dry seasons; of course, these climatic variations affect vegetation and groundwater storage as well as stream flow.

Besides the more or less random temporary changes, runoff is sometimes affected by permanent changes in the climate or in the drainage basin. In rare cases, headward erosion of the master stream in a drainage basin cuts through the drainage divide and "captures" part of the stream system in the neighboring basin. This can usually be detected by careful study of a topographic map of the drainage area. Sometimes, too, observation of the landscape (or map study) will reveal a broad valley that was eroded by a large volume of flow in the past but that now contains only a small stream in its central part. Stream flow obviously has diminished due to a much smaller water supply. A number of these oversized valleys can be seen in the northern United States. They were eroded by heavy flows of meltwater during the final recession of glacial ice about 12,000–15,000 years ago. The modern climate now supplies far less water for stream flow in these old valleys.

The best index of all the factors that affect runoff is a record of flow in the main stream draining a basin. Measurement and interpretation of stream flow is one of the chief occupations of hydrologists concerned with the study and control of surface water. Water supply, waste disposal, inland navigation, and flood control consume a sizeable proportion of the public works part of the federal budget, and all of these activities depend on accurate stream-flow measurements.

MEASUREMENT OF RUNOFF

When rain falls or snow melts, water runs everywhere over the surface of the earth, in sheets of overland flow, in myriad tiny rivulets, in small channels, and finally in streams and rivers. There are also unseen components of runoff beneath the surface in zones of interflow and in deeper zones of groundwater flow. Measuring all these components of runoff would seem to be an impossible task.

Fortunately, however, there is a practical solution to this dilemma. You don't have to measure all the upslope components of runoff. Since all of these come together downslope in the streams, by measuring stream flow you can measure runoff from the entire surface of the drainage basin.

It sounds easy, and in principle it is. As the discussion proceeds, however, it will become apparent that simply measuring stream flow is only the beginning. To accurately describe the runoff from a drainage basin, the environmental characteristics that affect runoff must be known and accounted for. How fast does rainfall enter the stream during a storm? What is the effect of antecedent precipitation in saturating the soil? What is the contribution of groundwater flow to the stream before, during, and after a storm? Interpretations of stream-flow records have generated a considerable body of hydrological and engineering literature. It all begins, however, with a technician called a hydrographer, going out with a current meter and actually measuring the stream discharge.

Measuring Stream Flow

The oldest known records of stream flow are those begun about 5,000 years ago by the Egyptians along the Nile. They used staff gages, called Nilometers, that they fastened permanently at the river's edge, allowing them to measure the rise and fall of river level. The purpose was to keep a record of river stage and to anticipate the arrival of the flood wave that sweeps down the Nile Valley each summer from the mountains of Ethiopia. Located at strategic places along the river, the gages were read by boatmen, who then rowed swiftly downstream ahead of the flood and alerted the farmers who were waiting for the water in the farmlands of the Nile delta. These Nilometers served the Egyptians well, as they had no particular need to know how much water was flowing. They just wanted a record of river stage and some warning as to when the flood waves were moving downstream.

In modern times, we have found it necessary to measure not only the water level, or *stage*, of a stream but its total flow volume as well. If you plan a reservoir on a stream to store water for a city supply, irrigation, or flood control, you need to know how much water the stream is likely to provide in a year or in a single big storm. When you wade across a stream or go across in a cable car and make current-meter measurements of the flowing water, you can't help but feel a direct involvement with the process of runoff. This is the most direct measurement you can make on any phase of the water cycle.

When the volume of stream flow is measured or reported, the term used is *discharge,* defined as *unit volume of water flowing in the stream per unit of time.* The two most common units of discharge are the cubic foot per second, used mainly in the United States, and the cubic meter per second, used throughout most of the rest of the world.

In principle, stream measurement is easy. You simply measure the area of a cross section of flowing water along a line across the stream at right angles to the direction of flow, then measure the velocity of the flowing water. Multiplying the area of the cross section by the velocity produces a figure for the instantaneous discharge at time of measurement. Sound confusing? A simplified model will show how it works.

Visualize a box with a square cross section, as in Figure 8.7. If the box is one foot wide and water flowing is one foot deep, the cross section of flow

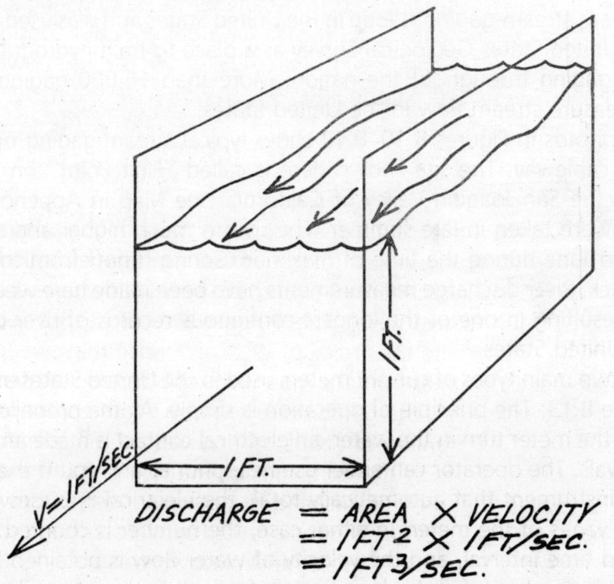

FIGURE 8.7 Discharge measurement in cubic feet per second.

has an area of one square foot. If the water is flowing at a velocity of one foot per second, the discharge, which is computed by multiplying the velocity by the cross-sectional area, is one cubic foot per second. (In metric measure, the discharge would be given in cubic meters per second.)

Measurement of discharge in a real stream is done along the same lines, with the hydrographer, or stream gager, making the appropriate measurements with a tape and a current meter. Where the water is shallow and not too swift, the hydrographer can select a place to measure and stretch a tape across the stream. The tape (or sometimes a wire with markers fixed at equal intervals along it) is anchored on each bank; the gager wades across, measuring water depth and current velocity at each interval marked on the tape or wire. A hydrographer provided Figure 8.8, which demonstrates how this is done.

Most measurements are made on larger streams in which the water is too swift or too deep for wading. Sometimes a bridge is located so it can be used by the stream gager; then it is a simple matter of walking across and lowering the current meter at regular intervals along the length of the bridge. Frequently, however, no bridge is located at the place chosen for measurement. In this case, it is standard practice to build a cableway across the stream and make measurements from that. Figure 8.9 shows the cableway across the Rio Grande at Embudo, New Mexico. This was the first permanent stream-gaging station in the United States and was used originally by the United States Geological Survey as a place to train hydrographers for stream gaging throughout the nation. More than 16,000 gaging stations now measure stream flow in the United States.

The photos in Figures 8.10–8.12 show typical stream-gaging operations from a cableway. The site shown here is called "First Point" on the Kern River in the San Joaquin Valley of California (see Map in Appendix I). The photos were taken in late summer. The river is much higher and swifter in May and June during the time of maximum spring runoff from the Sierran snowpack. River discharge measurements have been made here weekly since 1894, resulting in one of the longest continuous records of river discharge in the United States.

The two main types of current meters used in the United States are shown in Figure 8.13. The principle of operation is simple. As the propeller or little cups of the meter turn in the water, an electrical contact is made and broken at intervals. The operator can either use headphones and count the clicks or use an instrument that automatically totals the electrical impulses from the turning vanes of the meter. In either case, the number is counted during a standard time interval, and the velocity of water flow is obtained by multiplying this number (of clicks) by a "meter factor." Each meter is individually calibrated by the manufacturer, and each meter has its own multiplying

FIGURE 8.8 Stream discharge measured by wading.

FIGURE 8.9 Stream gaging station at Embudo, New Mexico, the first permanent gaging station in the United States.

factor. The calibration is carried out originally by drawing the meter through a tank of water at predetermined speeds.

In the field a stream measurement is made by first dropping the weighted meter to the bottom of the stream to determine water depth, then raising it to a predetermined fraction of the depth for the actual velocity measurement. The United States Geological Survey usually recommends making two velocity measurements, one at 80% and one at 20% of the water depth, then averaging these values to arrive at the average velocity of the stream at that place. Another common practice is to make one measurement at 60% of the total depth and to use this as the average stream velocity for

FIGURE 8.10 Hydrographer preparing to travel across stream on cableway. The handle hanging down from cable above the gager's head is used to pull the cable car "uphill" beyond the sag in the cable at midstream. The current meter has a heavy lead weight attached to make the instrument sink in the flowing water.

FIGURE 8.11 Lowering current meter in the river.

FIGURE 8.12 Preparing to take a reading at midstream.

FIGURE 8.13 Current meters in common use in the United States. The instrument on the left is one type of propeller meter called an Ott meter. On the right is the Price meter, the standard meter used by the United States Geological Survey.

that segment of the cross section. The reason for this is that the velocity of the stream is not uniform in a vertical section (see Figure 8.14). Figure 8.14 also shows why timing the downstream progress of a floating object will not give a completely accurate value for average velocity. The surface water moves slower than the deeper water in the stream.

By progressing across the stream and measuring depth and velocity at equal intervals, the stream gager ends up with a series of figures for average velocities in each segment of the cross section. The geometric pattern would look something like the sketch in Figure 8.15. Each velocity is multiplied by the area of its segment to find discharge through that segment, and the total of these added together gives the desired figure for stream discharge.

When the stream gager computes the discharge from the current-meter measurements, the result is the instantaneous discharge at that time and at that stream level. While this is absolutely essential information, it is not

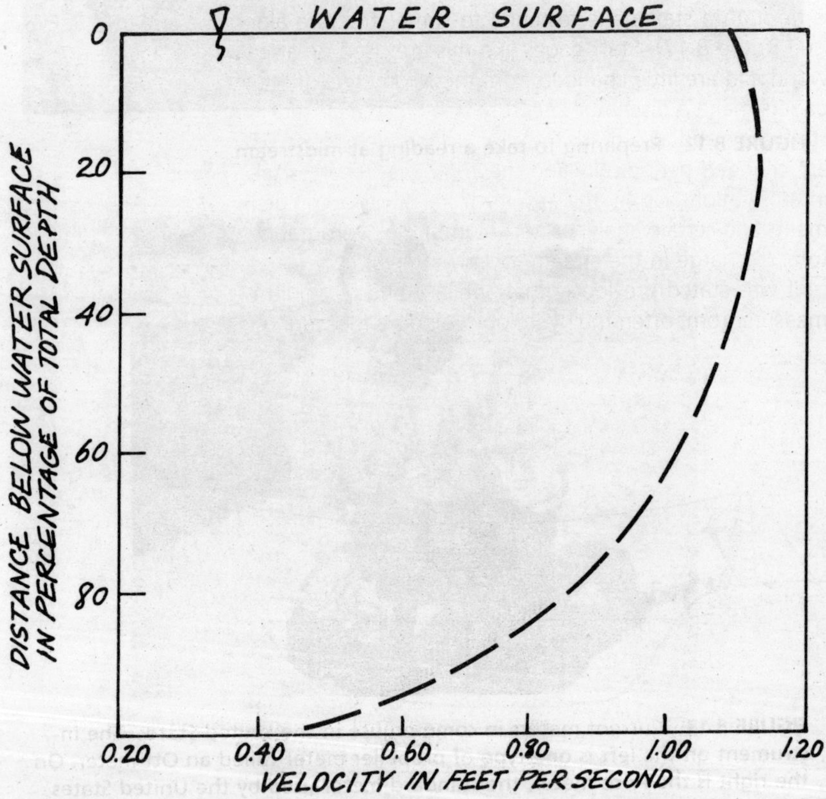

FIGURE 8.14 Typical vertical distribution of velocity in a stream.

enough to satisfy the needs of the engineers and planners. What is needed is a way to monitor discharge continuously so that stream-flow totals can be known for specified time intervals. How much water did the stream carry each day or each week last year? How much, on the average, for the past 10 years? How much flow was there from the storm last week?

Actual measurements with current meters at a gaging station occur infrequently—sometimes once a week, more often once a month, or even at longer intervals. The problem then is to determine discharge between the gager's visits. This is done with a continuous record of stream level, which is correlated with the occasional current-meter measurements. Looking back at Figure 8.9, you will notice a little stone house at the far right of the photo. You will see some kind of structure like that at most gaging stations (but most are not such fancy stone houses as this one!). These structures enclose an automatic water-level recorder that sits over a stilling well connected to the stream, as shown in Figure 8.16. Another way of checking water level is through a staff gage fastened to some structure along the stream bank (see Figure 8.17). Staff gages like this are used on streams throughout the world and are little changed from the Nilometers of 5,000 years ago. When gagers make a discharge measurement with a current meter, they read the staff gage and also mark the chart in the well house. These recorder charts are changed periodically and the completed chart is taken to the office for interpretation. Using the results from occasional current-meter measurements and certain analytical techniques, the hydrographer is able to compute total discharge in the stream for any intervals.

It was stated previously that, while a bridge would be a handy place to measure from, often the hydrologist wants a measuring station at a different

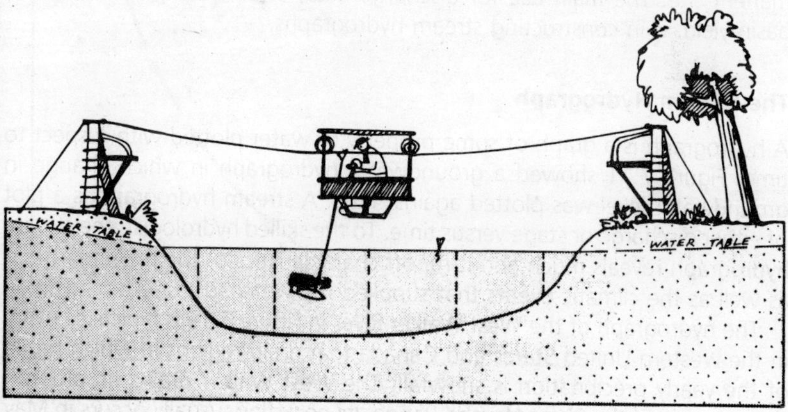

FIGURE 8.15 Sketch of stream cross section for measuring stream discharge.

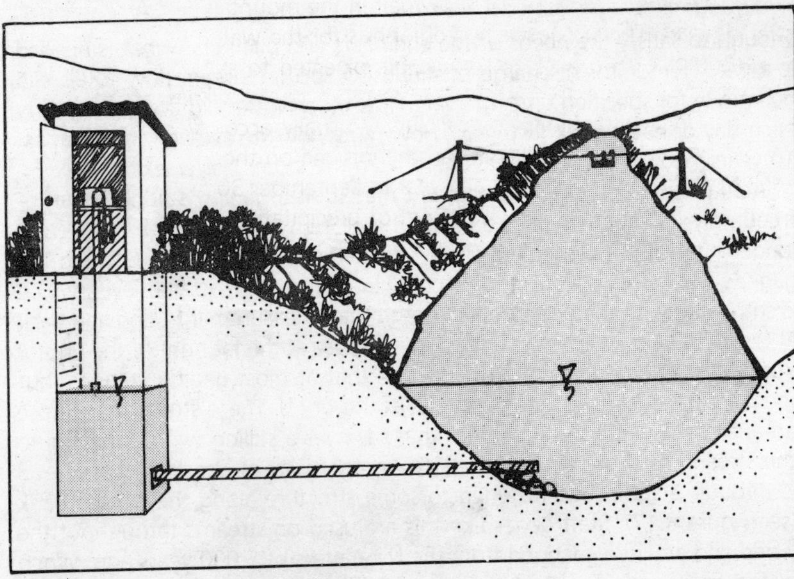

FIGURE 8.16 Stilling well with recorder for making a continuous record of stream level at a gaging station.

place. Why? It turns out that a number of conditions can signify a good or poor site for measuring streamflow. A technical discussion of the criteria for selecting a stream-gaging site is beyond the scope of this book, but for the record of discharge to be useful it should be a continuous record at a permanent site. The main use for discharge data, aside from gross totals for basin yield, is in constructing stream hydrographs.

The Stream Hydrograph

A hydrograph is a graph of some property of water plotted with respect to time. Figure 7.11 showed a groundwater hydrograph in which change in groundwater level was plotted against time. A stream hydrograph is a plot of either discharge or stage versus time. To the skilled hydrologist, the stream hydrograph reveals much about runoff characteristics of the drainage basin as well as the climatic events that supplied the water for runoff.

The hydrograph of the West Walker River in Figure 8.18 is typical of rivers in the western United States and Canada that drain regions in which much of the yearly precipitation is snowfall. The West Walker drains an area on the east side of the Sierra Nevada Range. Its peak flow usually occurs in May and June and is the result of snowmelt in the mountainous regions drained by the river. This particular hydrograph is for the water year 1981–82, but hydrographs for other years could be expected to show the same annual pattern.

The United States Geological Survey usually accounts for stream flow, groundwater storage, and similar phenomena on the basis of a *water year* that runs from October 1 of one year to September 30 of the next year. The reason for this is the annual pattern of precipitation and snowmelt, which

FIGURE 8.17 **Staff gage for reading water level in a large irrigation canal in California.**

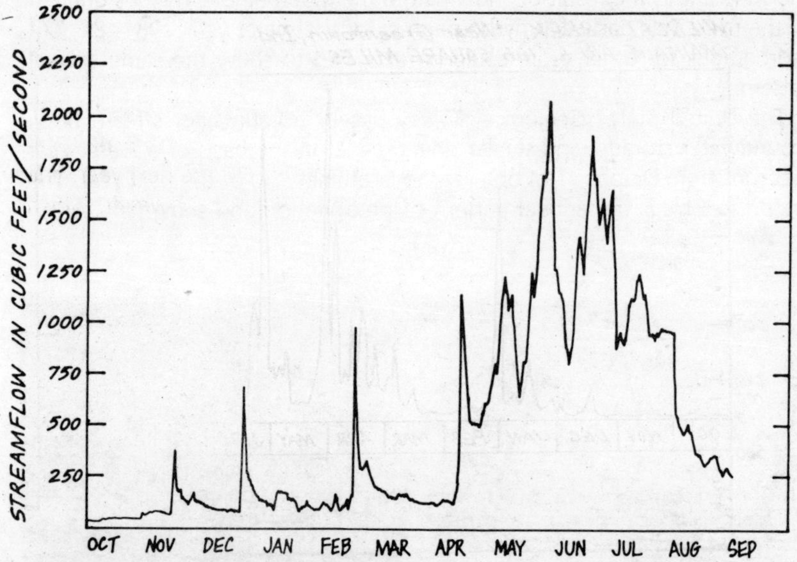

FIGURE 8.18 Hydrograph for the West Walker River, measured near Coleville, California, water year 1981–82 (source: U.S. Department of Agriculture, Soil Conservation Service).

generally translates into the lowest stream flows and groundwater levels each year during autumn, around October 1. This can be a little confusing because the National Weather Service reports precipitation on a calendar-year basis, January 1 to December 31. And in California, where very little rain falls in the summer, the department of water resources often reports precipitation for the period July 1 to June 30. Some reports of the department, however, list precipitation on the October 1 to September 30 yearly basis. The moral of all this is the time period for which stream flow, precipitation, and so on are reported must be carefully noted to make comparisons among regions or years.

Comparing the hydrographs for streams in two small basins in northern Indiana (see Figure 8.19) shows how much the soil and geology can affect runoff characteristics of drainage basins. The basins are roughly the same size and are subject to the same general pattern of annual precipitation. Because glacial deposits with a high clay content underlie Wildcat Creek basin, the surface soils have low infiltration capacity, and rainfall or snowmelt runs off rapidly. There is little groundwater storage, and most of the stream flow comes from storm runoff. In the basin of Tippecanoe River, on the other hand, the glacial deposits are mostly sands and gravels with high infiltration

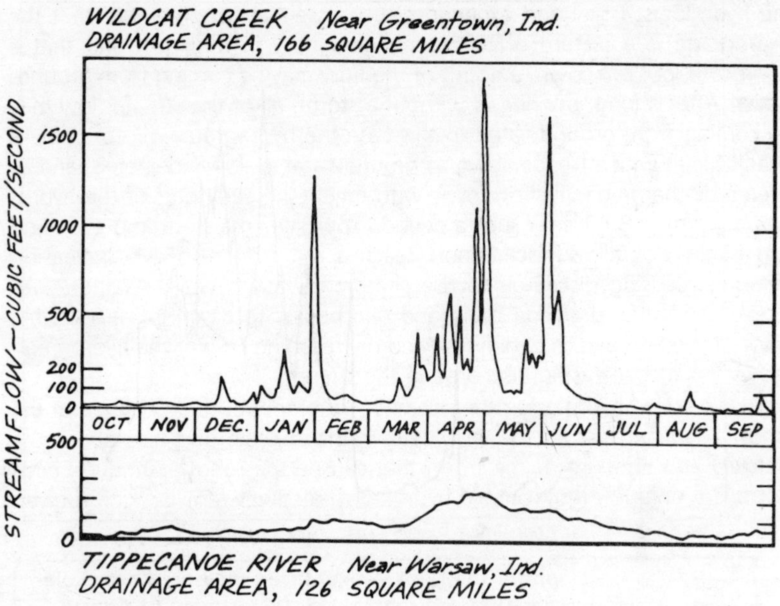

FIGURE 8.19 Hydrographs for streams in two drainage basins about 50 miles (80 km) apart in northern Indiana, water year 1946–47 (source: *Water*, Yearbook of Agriculture 1955. U.S. Department of Agriculture: 57).

capacities, and much of the rainfall and snowmelt seeps down to the water table. Most of the annual flow of the Tippecanoe, therefore, is base runoff from groundwater storage.

The three hydrographs shown in Figures 8.18 and 8.19 are useful for showing the general aspects of the drainage basins and the approximate volumes of annual runoff from the basins. The engineer concerned with controlling floods or designing bridges or water-control works, however, needs a more detailed picture of short-period storm runoff. Before building either a bridge across a stream or levees along a stream to protect property on the floodplain, you need to know how high and how fast the stream will rise under given storm conditions. For these and other reasons, hydrologists routinely construct hydrographs for brief time periods to study runoff response to storms of various lengths and intensities. A typical hydrograph showing stream flow from a single storm is pictured in Figure 8.20.

In analyzing a stream hydrograph, one of the first things the hydrologist does is to separate the hydrograph into its two main components: (1) base runoff from groundwater, and (2) direct runoff from storm rainfall (see Figure

8.20). The separation is done mainly on an empirical basis, depending on the hydrologist's skill and experience. It is essential to remember that the hydrograph is a picture of time-dependent flow in a stream and that a significant (but unknown) amount of the flow may be coming from groundwater. After a long, dry period without a storm, all of the stream flow may be coming from groundwater. In this case the hydrograph will resemble a straight line, which trends down as groundwater is slowly depleted, and as stream discharge diminishes slowly with time. (The early part of the hydrograph in Figure 8.20 is for such a period.) Then, when a storm moves in and rain begins to fall, surface runoff reaches the stream in ever-increasing amounts, causing discharge to rise sharply toward a peak. As the storm moves across the drainage basin and rain begins to diminish, at a certain point runoff also begins to diminish and the receding limb of the hydrograph shows this as decreasing discharge in the stream.

To the skilled hydrologist, the stream hydrograph reveals much about the runoff characteristics of the individual drainage basin and also about the intensity and duration of the storm that caused the runoff. Although each storm is a unique event, once a hydrologist becomes familiar with a given

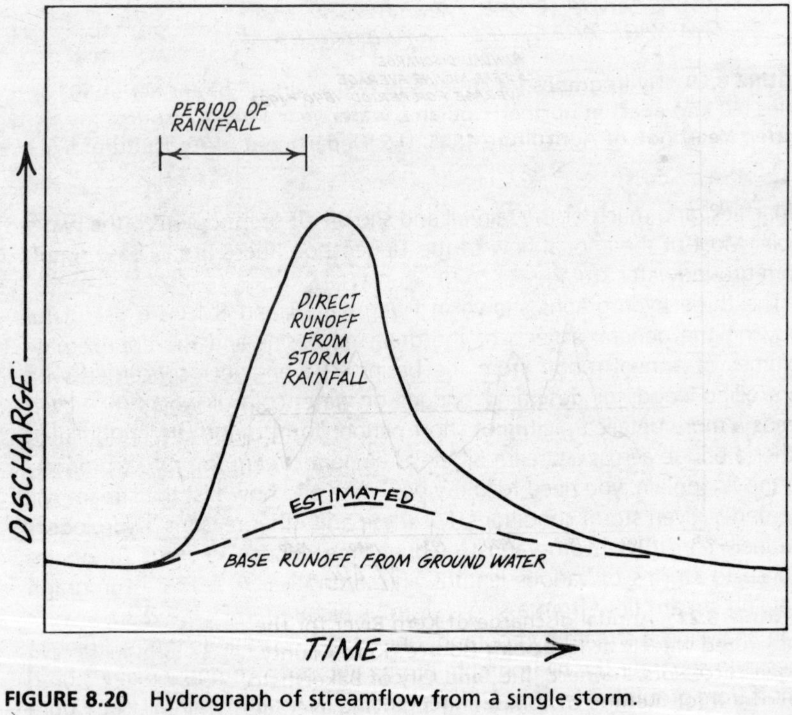

FIGURE 8.20 **Hydrograph of streamflow from a single storm.**

stream and its drainage basin, a recurring pattern often emerges, making hydrograph analysis easier and more reliable.

One type of hydrograph that hasn't been discussed is a long-term flow diagram made using annual totals of runoff, plotted by the year, for the entire period of record. Figure 8.21 shows such a diagram for the Kern River, measured at First Point (see Figures 8.10–8.12) in California's southern San Joaquin Valley. First Point is near the city of Bakersfield.

This chart reveals several interesting aspects of river flow as well as information on the climate in the southern Sierra Nevada Range, where most of the river's flow originates as snowmelt. The large differences in annual totals for river flow are striking. It is not uncommon for total flow one year to be far above the long-term average and for the following year to be below average. Without the five-year moving-average this is a real "scatter diagram" of data points. With it, though, one begins to see some order in all of the scattered annual totals. As the moving average deviates around the long-term average line, it is apparent that during several series of years, runoff was consistently above or below the average. That these periods were series of "wet" or "dry" years is also apparent when one compares the Kern River flow diagram with the long-term record of annual precipitation at

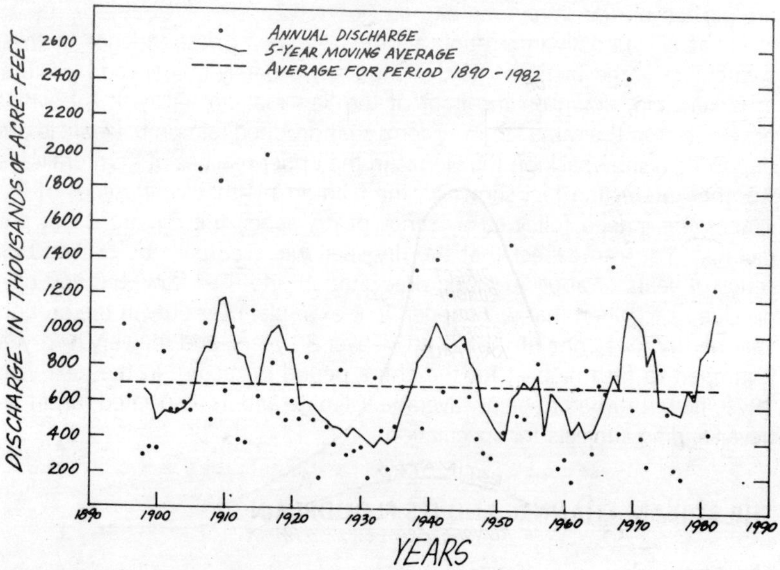

FIGURE 8.21 Annual discharge of Kern River for the period 1894–1982, measured weekly at first point (source: Kern County Land Company and their successors, Tenneco, Inc., and City of Bakersfield, California; (See Appendix II for method of computing a moving average).

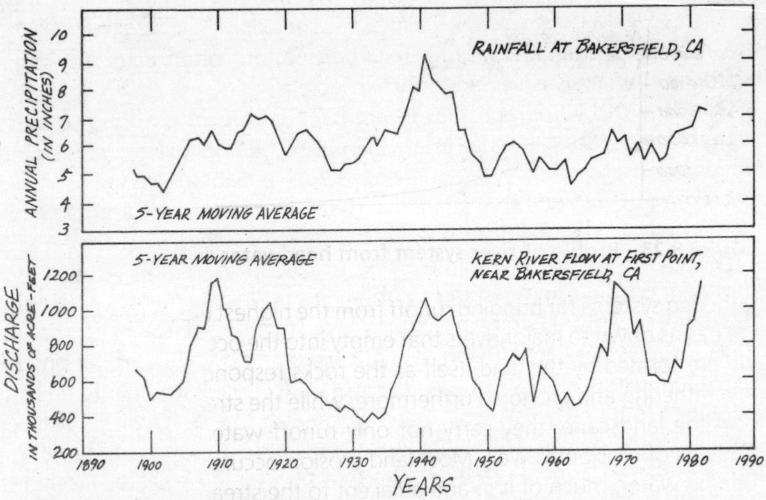

FIGURE 8.22 **Comparison of five-year moving averages for Kern River flow and rainfall at Bakersfield, California for the period 1894–1982.**

Bakersfield (shown in Figure 4.13). Moving-average lines from these diagrams are plotted on the same time base in Figure 8.22.

What is especially interesting about the close correspondence in these two curves is the fact that flow in the Kern is barely influenced by rainfall at Bakersfield, down on the floor of the San Joaquin Valley. It is obvious, however, that the same storm systems that dropped rain on Bakersfield also deposited a snowpack on the Sierras in the upper reaches of Kern drainage. Another interesting fact shown in the diagrams is that while series of wet years were indeed followed by series of dry years, the climate is not predictable. You can expect that the dry spell will eventually be broken by a series of years of above-average precipitation and river flow, but you have no clue as to when that will happen. For example, river flow in the drought year of 1977 was one of the lowest in over 80 years, and the very next year it jumped to fifth highest for the entire period of record. In the next year, 1979, flow returned to below average. Climate and its associated runoff are never-ending subjects for speculation.

THE STREAM CHANNEL AND ITS FLOODPLAIN

The Channel

The stream channel is the conduit through which runoff water travels over the land surface toward the sea. Networks of branching channels provide

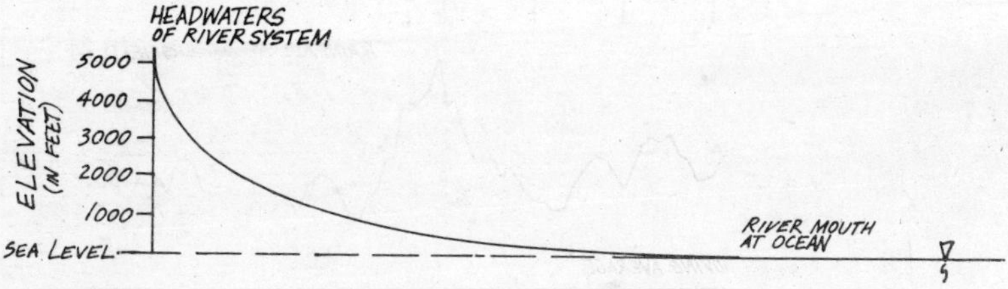

FIGURE 8.23 Profile of river system from headwater to the sea.

gathering systems for bringing runoff from the highest elevations of the drainage basins down to major rivers that empty into the ocean. The channels form and are formed by the land itself as the rocks respond to the natural forces of weathering and erosion. Furthermore, while the streams help to carve and form the landscape, they carry not only runoff water but the sedimentary products of erosion as well. Most land erosion occurs through the action of running water, much of it in and adjacent to the stream channels.

Natural drainage systems are dynamic, tending to adjust quickly to changed conditions of land slope or water supply. When plotted in cross section (along the length of the stream), natural stream gradients appear concave, reflecting the changing flow regimen as the stream goes from the highlands in the upper reaches of the drainage basin to the lowlands by the sea. Figure 8.23 is a typical profile of a river system from headwaters to the sea. Upstream, the stream is actively eroding its bed and often flows in a steep-walled canyon without a floodplain. The Green River in northwestern Colorado, shown in Figure 8.24, is such a river. Farther downstream where the gradient is flatter, the stream tends to erode laterally as well as vertically, and a floodplain appears on either side of the channel. Figure 8.25 shows a small Alaskan river meandering across a narrow floodplain where the stream gradient is fairly flat.

In arid regions, ephemeral streams often flow from mountain canyons onto broad alluvial plains on which the stream deposits sediment as its gradient flattens going from the mountains out onto the valley floor. Streams of this type flow only occasionally and often contain large volumes of water and sediment for short periods. Figure 8.26 shows such a stream valley, emerging from the eastern face of the San Jacinto Mountains south of Palm Springs, California (see Map in Appendix I). The residential subdivision shown on the lower end of this alluvial plain may disappear some day in a torrent of flood water and mud.

The Floodplain

The floodplain and its relation to the stream channel is a subject of considerable interest in the field of applied hydrology. That the floodplain is an

FIGURE 8.24 Whirlpool Canyon on the Green River in northwestern Colorado.

integral part of the stream itself is apparent in the definition of a floodplain. Although definitions vary somewhat among the different authorities on the subject, a definition that most would agree with goes something like this: A floodplain is a strip of relatively smooth land bordering a stream, built of

FIGURE 8.25 A small Alaskan stream meandering across its floodplain.

sediment carried by the stream, and overflowed regularly in times of high water.

With a little imagination, one can imagine the way a stream builds its floodplain by studying the photograph in Figure 8.25. As the water flows along the sinuous channel, it tends to erode the bank on the outside of the meander bends and to deposit the eroded material on the shallow bars (called *point bars*) on the inside of the bends. In the photograph the point bars show up as white areas on the inside of the meander bends. This process of erosion and deposition tends to cause the meander belt of the stream to move laterally, and in time it will swing from one side of the valley to the other, building a floodplain in the process. Some floodplain material is also deposited in times of flood when high water rises above the stream banks and overflows the flood plain. This is of minor importance in building the floodplain, however, when compared with the continuous work of erosion and deposition within the channel itself.

One thing that does change during a flood is the stream's erosive power. Larger volumes of water and higher velocities of stream flow during floods are responsible for much of the major erosion that occurs along stream courses. The stream may even displace itself and begin to flow in a new channel during a flood. This commonly results from breakthroughs between river bends and causes the stream to abandon meanders in its previous channel. Two such abandoned meanders can be seen on the floodplain of the stream in Figure 8.25. These usually persist for a while as lakes, called *oxbow lakes,* and they are common to many floodplains.

FIGURE 8.26 Mouth of Desert Canyon in east face of San Jacinto Mountains, California.

FLOODS

For descriptive purposes, we may talk about the *stream channel* and the *floodplain,* but these are both part of a *stream system* that nature has constructed to carry runoff water overland to the sea. In truth, the floodplain is simply the high-water phase and the so-called channel is the low-water phase of this flow system.

Usually, a perennial stream flows within the banks of its regular channel. In fact the channel is formed by and for these moderate flows, which are the most frequent flows during the year. Occasionally, during heavy storms, the water rises to the top of the banks. These flows occur about twice a year in most streams in humid regions, and the channel can just barely contain them. During heavier storms or prolonged periods of snowmelt, the water may even spill over the banks and onto the floodplain. Overbank flows are true floods, with the floodplain then temporarily becoming part of the stream channel. Most perennial streams have an overbank flow about once every two years on the average.

The amount of water flowing in a stream varies during the year—and even from day to day—although the variation in flow receives little attention (except from the stream gager) until it spills out over the bank. Then it gets plenty of attention. Most people, including the news media, always seem surprised when a flood occurs. But to a hydrologist the overbank flow of a flood is just as expected as other levels of flow. Besides level, the only difference among the various flow volumes is their frequency. Hydrologists can't pinpoint exactly when a particular volume of flow will occur, but they know from the analysis of stream flows in hundreds of streams over many years that floods are an inevitable part of runoff in most places on the globe. Figure 8.27, adapted from the U.S. Geological Survey's "Primer on Water," shows this concept graphically.

For the hydrologist, predicting floods is just a matter of figuring the odds. If you've ever played poker, you can understand flood frequency in terms of the kinds of hands you are dealt in poker. Stretching the analogy a bit, and looking at Figure 8.27, you could say that A is typical of a poker hand with nothing much in it (no pairs, etc.), B would correspond to a hand containing one pair, C might correspond to 3 of a kind, D to a flush (all cards one suit), E to a straight flush (cards of one suit in continuous sequence) and the 100-year flood (not shown) to an ace-high straight flush, or royal flush.

Such an analogy may sound frivolous, but there's actually more there than you might think at first. One thing that poker players know—something that makes the game so fascinating to many players—is that while the odds of getting a royal flush are very long indeed, it is still possible to be dealt such a hand twice in rapid succession. Even though the laws of probability state that such an event will occur only once, on the average, in 650,000 hands, the cards have no memory and an unlikely event can occur at any time in a game. (That assumes an honest dealer, of course.) Nature plays the "runoff" game in the same way. Hydrologists speak of the "recurrence interval" of various flood flows, and the so-called 100-year flood can occur (and has occurred) twice within just a few years. A point to remember here is that, according to the laws of probability, if you can assign a probability to the occurrence of an event, given enough time, no matter how small the probability, the event will eventually take place.

Flood Control

The subject of floods is of intense interest in many riverside communities. Most people don't understand why streams can't manage to stay in their channels when high water comes. To many it seems that nature is playing "dirty tricks" when floodwaters cover their neighborhood and leave deposits of mud in their houses. Newspaper headlines speak of a "river on the rampage." State governors fly over the flooded areas in helicopters and declare a "state of emergency." If the flooded area is large enough, the President may even declare a "state of emergency." These declarations of emergency states allow people in the flooded areas to apply for low-interest, govern-

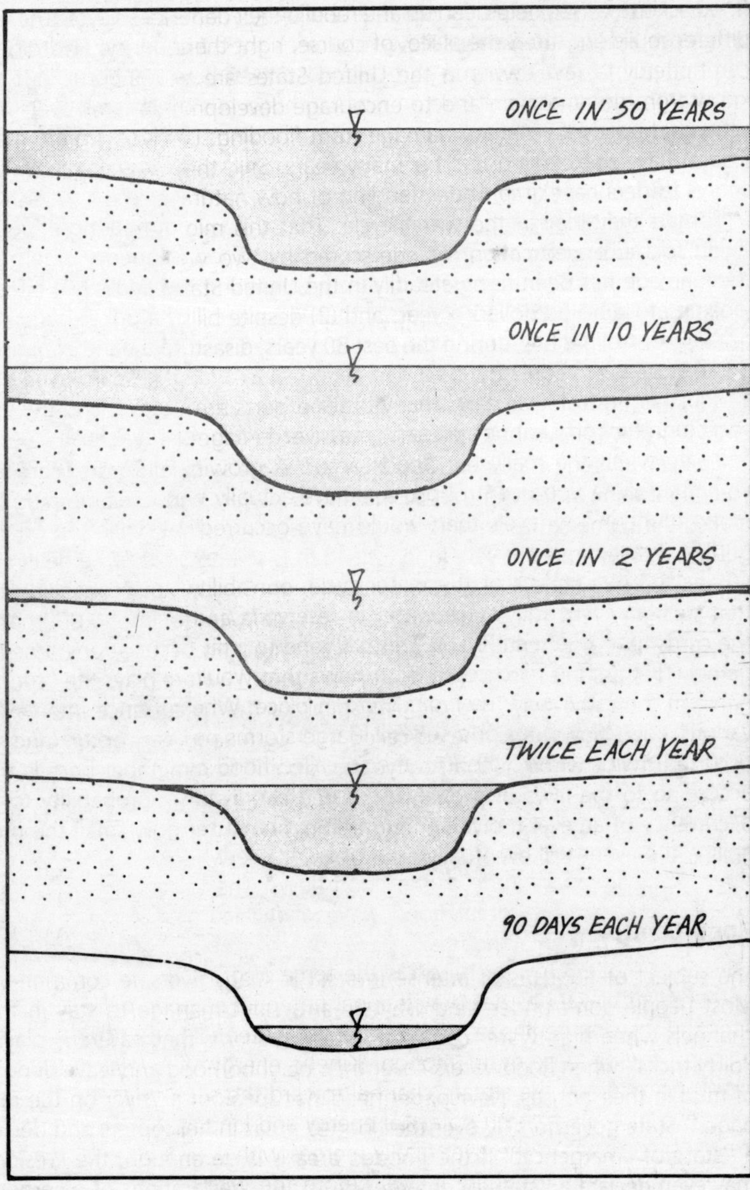

FIGURE 8.27 **Levels of flow in a typical stream channel (source: U.S. Geological Survey).**

ment-backed loans to help clean up and rebuild their damaged homes. Where do they rebuild? In the same place, of course, right there on the floodplain.

Admittedly, a few towns in the United States are beginning to restrict land use on the floodplain and to encourage developments, such as parks, that would not sustain heavy damage from flooding. Local governments in England have been doing this for many years. Still, the general public and political leaders have little understanding of how natural streams work and of the part they play in the water cycle. That this might be an important subject for public education is underscored by two well-known facts: (1) flood damage has been rising steadily in the United States since 1900, now exceeding four billion dollars a year, and (2) despite billions of dollars spent on flood-control projects during the past 80 years, disastrous floods still occur with discouraging regularity.

Engineers, hydrologists, and many other experts are well aware that the flood-control efforts cannot prevent massive damage to property on the floodplain when the really big floods occur. A growing body of evidence shows that some flood-control projects have actually worsened conditions, causing higher flood levels than would have occurred without the control structures.

As with other aspects of the water cycle, our ability to control running water is limited. We may store water in reservoirs and attempt to confine the river channel between levees. This will tend to limit property damage on the floodplain during moderately high flows that would ordinarily spill over the banks if no levees or reservoirs were in place. When a large snowpack begins to melt all at once or a series of large storms persists over upstream regions of a river system, though, the so-called flood control structures are apt to be overwhelmed by rising water. Then the only thing people in downstream communities can do is take to boats or head for the hills. At such a time even the established gaging stations are overwhelmed and one can only estimate the magnitude of the flood flows. The total energy involved in the passage of these great flood waves is incalculable, but it must be tremendous by human standards.

The source of energy for runoff is the force of gravity. Water falling on land at higher elevations has gravitational potential energy—energy of position. As the water begins to run downhill, the potential energy is converted to kinetic energy—energy of motion. Kinetic energy is what propels the running water down the stream channel toward the sea. When the stream finally reaches the sea, both potential energy and kinetic energy have been expended. As water in the river finally merges with water in the ocean, its energy state returns to where it was before the water was picked up by evaporation and started on its long journey. The circle is complete. As Ecclesiastes (1:7) noted, ". . . unto the place from whence the rivers come, thither they return again."

APPLICATIONS

Applications for the runoff part of the water cycle fall into two main categories: *water supply* and *flood control.* Most of the water we use, for any purpose, comes from surface runoff through the rivers and streams and their associated reservoirs. Surface runoff is also the source of floodwater. Probably the most comprehensive application to consider would be one that included aspects of both water supply and flood control. If that were the case, then an application involving a whole river system might be the way to go. The Colorado River System is one of the best for this purpose.

The Colorado River Basin[2]

The Colorado, the major river of the Southwest, was first seen by white people when Spanish conquistadores discovered it in 1540. Prehistoric Indians had been living and hunting along this great river for several thousand years before that. Both the Indians and Spaniards viewed the river as a natural phenomenon that one lived with and accommodated to in all its modes, whether in flood or in drought. The colonists who began to come into the country after the middle of the nineteenth century, however, had a different view of rivers and of nature in general. These settlers were practical. They believed in the scriptural dictum that our mission on the earth was to subdue and control nature for our own benefit. Where the Spaniards brought missionaries to convert and subdue the Indians, the colonists brought engineers to subdue and control the rivers.

Controlling the River. Within 20 years of John Wesley Powell's pioneering explorations through the Colorado's canyons in 1869, engineers of the U.S. Geological Survey established their first river gaging station on the Rio Grande at Embudo, New Mexico (Figure 8.9). Shortly after, the survey engineers began measuring river flow throughout the West to determine the potential supply available for irrigation and other uses.

After the U.S. Bureau of Reclamation (USBR) was established in 1903, it was only a matter of time until plans were being made to tame the western rivers by building dams and reservoirs to control floods and provide water for irrigation. The first large dam was Roosevelt Dam, completed in 1911 on the Salt River in Arizona. And less than 26 years later, in 1935, Lake Mead began to fill behind Hoover Dam on the Colorado. Other dams followed in rapid succession until the major rivers, and even many tributary streams, were at least partially controlled by dams and storage reservoirs. Because it

[2]In preparing this section on the Colorado River Basin, the author used data supplied by the U.S. Bureau of Reclamation and Professor L. J. Paulson, Director of the Lake Mead Limnological Research Center at the University of Nevada, Las Vegas.

is the largest and most-used stream in the Southwest, the Colorado is the most controlled of all.

The Colorado and its major tributary, the Green, rise in the Rocky Mountains of Wyoming and Colorado and flow more than 1400 miles (2550 km) through parts of seven western states (see Figure 8.28) before finally emptying into the Gulf of California in Mexico. The drainage basin covers about 245,000 square miles (635,000 km²) or about 8% of the area of the conterminous United States. The long-term average annual runoff is estimated to be about 15 million acre-feet (18,502,000,000 m³), about 1% of the total annual runoff from the conterminous United States. It is easy to see why this river, draining about 1/12 of the land area but carrying only about 1/100 the runoff, is the most-used river in America. It might even be the most-used river in the world.

Dividing the Water. Despite needs for water within the river basin itself, more water is exported out of this basin than in any other river system in the country. Exports began in the 1890s, when water was diverted from the Colorado's headwaters to the east slope of the Rockies and into the South Platte River drainage near Fort Collins, Colorado. Now diversions take additional water to the Denver metropolitan area, into the Upper Rio Grande drainage in New Mexico, to the Salt Lake City metropolitan area, and by means of the Colorado River Aqueduct to the service area of the Southern California Metropolitan Water District.

One reason these diversions could be made was simply a lack of population in the basin (and hence of water demand), particularly in the upper basin. Another reason was a serious error by early-day engineers regarding the average flow to be expected in the river. In 1922 the seven basin states negotiated the Colorado River Compact, which divided the waters of the Colorado River system between the upper and lower basins at Lees Ferry (see Figure 8.28). At that time, existing records (1896–1921) indicated that the average annual discharge at Lees Ferry was at least 16.8 million acre-feet (20,723,000,000 m³). Subsequent measurements have shown that flow during 1896–1921 period was well above the long-term average, now considered to be about 15 million acre-feet (18,502,000,000 m³) per year at Lees Ferry.

The 1922 compact divided the water between the upper- and lower-basin states so that the lower basin would get 75 million acre-feet (92,512,000,000 m³) per decade (an average of 7.5 million per year) at Lees Ferry. The lower basin is also allowed to increase its use by an additional 1 million acre-feet (1,233,000,000 m³) per year. No mention was made of Mexico's share.

Mexico's allotment was not fixed until a treaty was finally negotiated in 1944. By that time no one really wanted to give Mexico very much water, but the Mexicans had a strong bargaining position. Most of the tributary

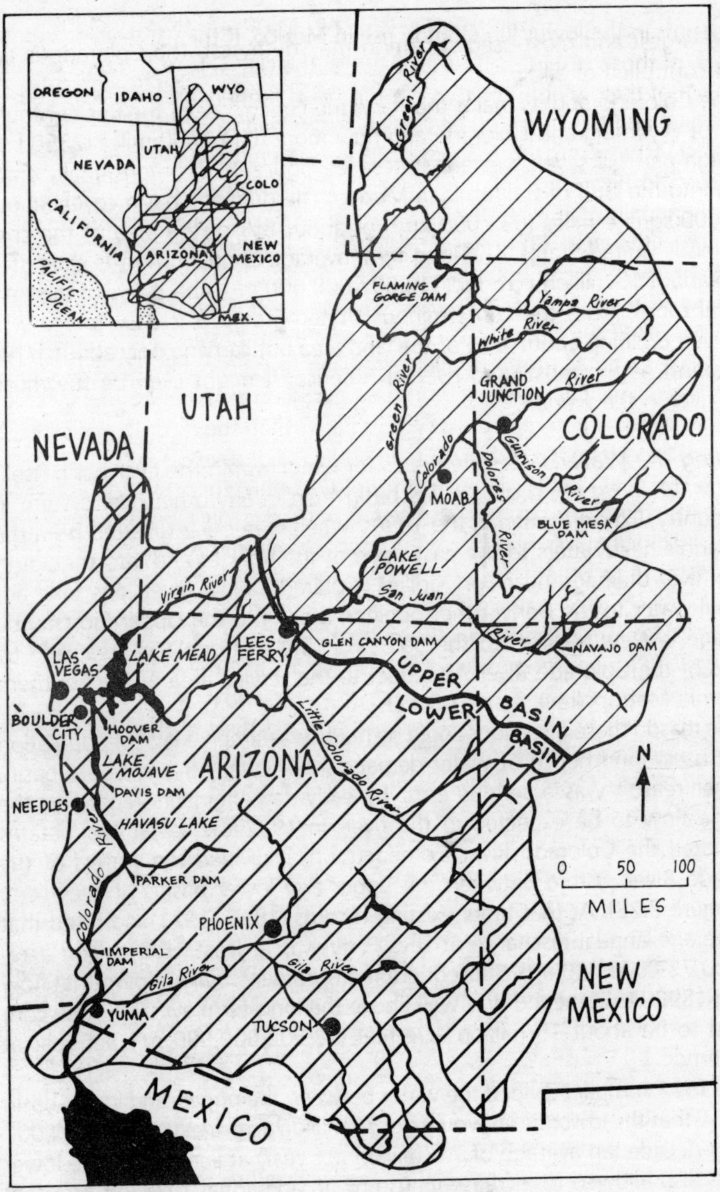

FIGURE 8.28 Colorado River Basin showing division into upper and lower basins at Lees Ferry.

streams in the lower Rio Grande rise in Mexico. If the Mexicans diverted the flow of those streams, citrus growers on the U.S. side of the river would lose much of their irrigation water. So the Texas farmers lobbied hard in the U.S. Senate, and the Colorado River water users had to give in. The treaty signed in 1944 between the U.S. and Mexico covered both the Rio Grande and Colorado Rivers and allotted Mexico 1.5 million acre-feet (1,850,000,000 m^3) annually at the border.

Over the years other agreements and court decisions have governed use of the river—most notably the Supreme Court decision of 1963 dividing water in the lower basin between Arizona and California. Collectively all of this is called the "Law of the River," and it has provided a very comfortable living for a small group of Western water lawyers.

If you have been adding up the various allotments, you already know that more water is potentially allotted on paper than the river can deliver in most years. Long-term, carry-over storage in the major reservoirs has helped smooth out the peaks and valleys in the water-supply curve, but as you will see shortly, there are costs in lost water, and reservoir storage is not the final solution. What has really saved the situation from becoming desperate until now is the fact that the upper basin states and Arizona have not been using their full share of water. When the Central Arizona Project comes on stream in the late 1980s, it will take a substantial amount of lower basin water and may result in water shortages in Southern California. Several Indian tribes are also claiming rights to water from the river, and if they win in the Supreme Court (or in Congress) the water they get obviously must come out of some area's present allotment. Looking back at the drafters of the 1922 Colorado River Compact, one could paraphrase Josh Billings: "It would be better not to promise so much water than to promise water that ain't there."

Stretching the Supply through Storage. Storage reservoirs don't increase the total water supply, but they can increase its usability by stretching out the period of use. Short-term storage is widely used in urban systems to provide water for periods of peak demand during the day or season. The neighborhood water tank is a familiar sight, and city reservoirs are common throughout the country (e.g., Lake Heffner for Oklahoma City). Long-term storage, on the other hand, has a different purpose and is not used as widely as short-term storage.

Long-term storage, also called carry-over storage, is used not so much to meet peak demands, but rather to save peak flows in a river system and thereby to stabilize the supply for continuous use during periods of one to several years. Locations that might require long-term storage are suggested by the pattern of annual precipitation and runoff throughout the conterminous United States, as shown in Figure 4.12.

In the more humid eastern half of the country, abundant precipitation throughout the year generally provides plenty of water in all seasons. Over the past several decades, only a few widespread disruptions in water supply due to drought have occurred. Short-term storage in relatively small reservoirs has been adequate to meet most water demands.

Conditions in the more arid western states create a completely different water-supply situation. Rainfall and snowfall are seasonal; without carry-over storage on the major streams, a large part of the annual supply would be unavailable for use. In much of the West, spring snowmelt produces most of the annual runoff. Without a provision for carry-over storage, most of the spring freshet would flow out to sea and be lost. Carry-over storage saves the snowmelt for use throughout the summer and makes this dry country habitable.

All of this is well illustrated in the Colorado River Basin. The river, often a raging torrent during the time of spring snowmelt, just as often became little more than a muddy trickle in late autumn before the storage dams were built. Now releases from storage reservoirs located throughout the basin (see Figure 8.28) keep the river flowing at a respectable rate all year long.

All the major reservoirs in the Colorado River basin are multiple-purpose reservoirs. They are operated by the USBR for the benefit of domestic use, irrigation, hydroelectric power generation, water quality control, fish and wildlife, recreation, flood control, and to meet all requirements of the Colorado River Compact and subsequent court decisions and treaty obligations. With all these competing needs, the frequent arguments among water users are not surprising, especially during years when basin runoff is very far above or below normal. Some users think the river should have even more storage capacity along its route than it does now.

The present capacity of the major reservoirs is a little more than 6l million acre-feet (75,243,000,000 m^3). With an average annual river flow of about 15 million acre-feet (18,502,000,000 m^3), this is a ratio of about 4 to 1 for storage capacity to annual flow. No other U.S. river system even approaches this ratio of storage to flow. Also, it is unlikely that additional storage along the Colorado River would save more water in the long run; in fact, the reverse is more likely. More storage probably would reduce long-term supplies because of the uncontrollable losses that go along with water storage in this dry climate.

Water Lost from Storage. Under natural conditions, every river system loses a certain amount of water to evaporation. Storing water in reservoirs behind dams tends to increase evaporation losses for several reasons. The area of water surface exposed to the atmosphere is larger than in the natural river, and the more or less static body of water in the lake behind the dam accumulates much more energy from solar radiation than does water in a free-flowing river. The operation of most dams and reservoirs also promotes accumulation of heat energy in the surface waters. Large dams generally have hydroelectric plants at their base, and most of the water released from the reservoir under normal operations is passed through the turbines to generate electric power. Since the power-plant intake is usually in deep water at the upstream face of the dam, the water released from the reservoir is almost always from the cold-water zone (called the *hypolimnion*) of the lake. This tends to maximize energy storage in the surface water, promoting evaporation from the reservoir. In the Colorado River Basin the major reservoirs lose more than 1.5 million acre-feet (1,850,000,000 m^3) per year to evaporation.

Another source of water loss is bank storage in the walls of the reservoir. Depending on geology of the reservoir site, this loss can be great or small. In most reservoirs, the water that enters walls or banks when water levels rise leave again soon after water levels fall to a lower elevation. For the normal operational range of water levels in most reservoirs, then, bank storage merely causes a time lag in the passage of water through the active zone of the reservoir. If a prolonged drought significantly lowered the water level, presumably the bank storage in the deeper levels of the reservoirs would also run out into the free water body fairly quickly. This is what most reservoir operators assume, and most of the time it is probably what actually happens. For example, the USBR engineers who operate Hoover Dam and Lake Mead assume that bank storage makes up about 6.5% of surface storage and they assume a 100% recovery of bank storage when reservoir levels drop.

As you have probably already surmised, all this talk of "guessing," "estimating," and "assuming" means that bank storage is one more hydrologic variable that cannot be measured directly. It is what remains after the other items in the water budget are either measured or calculated for a specified period. These other items are inflow, outflow, changes in reservoir storage, and evaporation. Thus, the residual that is termed bank storage includes any errors made in calculating the other terms of the water budget.

This is not an ideal situation, but it is the best we can do. Over the short run, the estimates may be faulty, but over a period of several years the errors probably balance out, giving a reasonable estimate for long-term bank storage. Unfortunately, however, we have no reliable method for predicting in advance the bank storage of an unbuilt reservoir. If projected bank storage is small relative to river flow and reservoir capacity, accuracy of the estimate doesn't much matter, especially if the water is rapidly recovered following a water-level decline in the reservoir. If, on the other hand, after the reservoir is built bank storage turns out to be larger than expected and a substantial amount of that storage is retained at low-water stages, the reservoir may not fill as quickly as originally projected; and some river water may also be permanently lost in the banks. This probably doesn't happen very often, but its occasional occurrence is shown by the history of Lake Powell behind Glen Canyon Dam.

Bank Storage at Lake Powell. Glen Canyon Dam and the lower reaches of Lake Powell are shown in Figure 8.29. When full, the lake extends for 186 miles (300 km) upstream from the dam. The dam was completed and Lake Powell began to fill in 1963. The lake did not reach its maximum elevation (i.e., it did not fill completely) until 1983, as seen in Figure 8.29. Why did it take 20 years to fill this second-largest storage reservoir on the Colorado? As you may already have surmised, the slow filling of the reservoir was caused by a combination of both hydrologic and geologic conditions, some of which were unusual and largely unpredictable.

FIGURE 8.29 Glen Canyon Dam and Lake Powell on the Colorado River. Photograph was taken in July 1983 when reservoir was full for the first time since it began filling in 1963. Overflow, being discharged through the spillway tunnel, is shown entering river at the lower right of photograph. Discharge from reservoir normally passes through power house, which can be seen just downstream of the dam. Both abutments of the dam are anchored in Navajo sandstone (photograph by U.S. Bureau of Reclamation).

As the major storage reservoir in the upper basin of the Colorado, Lake Powell is the ultimate source for the water that must be released to the lower basin under terms of the 1922 Colorado River Compact. Regardless of inflow, if the reservoir contains water, the compact releases must be made to the lower river.

The USBR engineers were well aware of the hydrologic uncertainties in projecting future levels in the reservoir. With a record of river flow spanning more than half a century, however, the engineers had at least a fair idea of how much inflow to expect at the upper end of the reservoir during an average year. The compact told them what the minimum outflow should be. Water losses would be due mainly to evaporation and bank storage. As outlined in previous chapters, expected evaporation from the reservoir could be estimated in advance with reasonable accuracy. Bank storage was the major unknown, and much thought and effort went into an attempt at quantifying the loss to be expected. That the attempt proved unsuccessful does not discredit the engineers' efforts. The Glen Canyon reservoir site is unique. Geologic conditions here are not duplicated at any other large U.S. reservoir, and perhaps at no other reservoir of comparable size anywhere else on the earth.

The walls of Glen Canyon are eroded in a massive red sandstone formation, referred to by geologists as Navajo sandstone. This is the same formation that forms much of the spectacular scenery in Zion National Park, about 100 miles (160 km) west of Glen Canyon. The Navajo is characterized by its uniformity and similarity in appearance over a wide area. Samples collected from outcrops 400 miles (644 km) apart are almost identical. They all consist of red to pink, well sorted, medium- to fine-grained, poorly cemented sandstone with an average porosity of about 25%. In Figure 8.29, the canyon walls and all of the rock in the foreground of the photo are composed of Navajo sandstone. This same material also forms the walls of the reservoir for more than 100 miles (160 km) upstream from the dam. Furthermore, the Navajo probably holds at least 80% of the water now in bank storage around Lake Powell.

Investigations by both the U.S. Geological Survey and the Bureau of Reclamation prior to construction of Glen Canyon Dam indicated that Navajo sandstone is permeable to groundwater flow. In its virgin state, the so-called glens of Glen Canyon contained numerous springs from which groundwater was seeping out of the canyon walls. When the dam's completion first caused levels in Lake Powell to begin rising, water began to flow in a reverse direction, from the lake into the walls of the canyon, and bank storage began to accumulate. The situation is diagrammed in Figure 8.30, which shows a typical cross section through the Navajo sandstone perpendicular to the canyon.

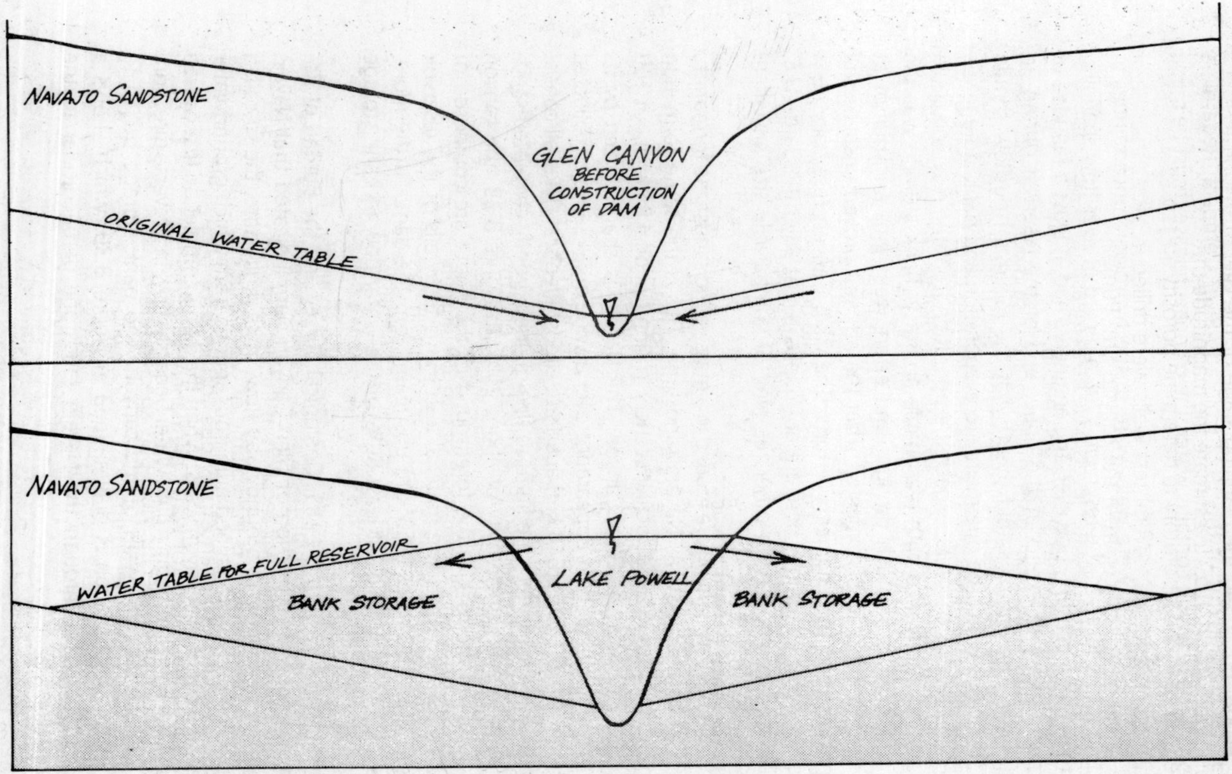

FIGURE 8.30 Generalized cross section through Navajo sandstone drawn perpendicular to Glen Canyon. The original water table and ultimate extent of bank storage are shown diagrammatically.

The potential for large water losses due to bank storage in the Navajo was recognized during the planning stage of the Glen Canyon project, and studies were made to try to estimate the magnitude of future losses. Samples were studied in the laboratory to determine porosity and permeability of the sandstone. In addition, several test wells were drilled in the sandstone bedrock at various distances from the canyon rim. Pumping tests were run on these wells, and groundwater levels have been measured periodically in some of the wells since before construction began.

With information from both the laboratory studies and field tests, the USBR engineers estimated in 1955 (before construction began) that bank storage might ultimately reach 2 million acre-feet (2,467,000,000 m^3). The engineers also assumed full recovery of bank storage when reservoir levels were drawn down. In 1971, eight years after Lake Powell began to fill, the USBR reported cumulative bank storage to be in excess of 6 million acre-feet (7,401,000,000 m^3). Later that same year a USBR report stated that while the ultimate volume of bank storage could not be determined precisely, it would probably be between 7 and 8 million acre-feet (8,634,500,000 m^3—9,868,000,000 m^3).

Actual losses to bank storage have been much larger than the projections. When water levels in Lake Powell finally reached maximum elevation in 1983 (signifying a full reservoir), the USBR estimated that bank storage was about 11 million acre-feet (13,569,000,000 m^3) and increasing at the rate of about 500,000 acre-feet (616,750,000 m^3) per year. It now seems obvious that the only reliable estimates of bank storage are ones made after the fact. The surface area of wall rock in contact with reservoir water and the total volume of pore space in the sandstone are both so enormous that the little data from laboratory studies and well tests made 30 years ago are totally inadequate to make an accurate calculation of ultimate bank storage. When water finally stops going into the banks, we will know they are full.

What about the estimates of recovery when reservoir levels fall? To answer that question with any precision you would need to know, among other things, the average specific yield and specific retention for the Navajo sandstone, as well as the original moisture content in the unsaturated zone above the water table. As you will recall from the definitions given in Figure 5.5 of Chapter 5, *specific yield* is the volume of water that will drain freely from a porous medium (e.g., sandstone) due to the force of gravity, and *specific retention* is the volume of water permanently held in the pores by capillary forces. The sum of specific yield plus specific retention equals the porosity, in other words, the total volume in the rock available for storing water. Since we don't know what the moisture content of the sandstone was above the original water table (Figure 8.30), we don't know how much of the water that went into the rock as bank storage was required to make up the specific-retention fraction of the total pore space. Hence we really don't have a basis for estimating how much of that water might eventually come back as specific yield and how much will be held permanently in the rock by capillary forces.

The data available on pore dimensions (from USBR laboratory studies) indicate an average pore diameter of about 0.03 millimeter (0.001 in.). If that pore diameter is representative of the sandstone, and if the pores were fairly dry when water entered the rock as bank storage, then it would seem reasonable to assume that a large part of the bank storage may now be held irretrievably as specific retention in the sandstone. The preproject assumption of full (100%) recovery of bank storage now seems implausible. The difficulty in estimating in advance the ultimate volume of bank storage itself should make us wary of estimating ultimate recovery. As in other instances where humans have disturbed a natural system, only the passage of time will reveal how much water nature is willing to release from the flooded walls of Glen Canyon.

Saving Water by Reducing Downstream Evaporation Losses. Although bank storage at Glen Canyon has taken, and probably will continue to take, water that would otherwise flow into Lake Mead, the way Glen Canyon dam is operated has also added to the lake's net storage. Except for years of high river flows (as in 1983, see Figure 8.29), the spillway tunnels at Glen Canyon are closed and water is released only through the power plant. This water normally comes from the deep, cold-water zone in Lake Powell. Since Glen Canyon Dam went into service, the river flowing through Grand Canyon and into Lake Mead has been much cooler than it was previously. The cooler water has resulted in less evaporation and in a net gain to the storage in Lake Mead. Prior to construction of Glen Canyon Dam, the evaporation at Lake Mead was estimated to average about 85 inches (2160 mm) per year. After normal river releases from Lake Powell began in 1964, the annual evaporation from Lake Mead decreased to less than 80 inches (2032 mm), and this reduction in water loss has increased storage in Lake Mead by an average of around a 100,000 acre-feet (123,350,000 m^3) per year.

Meeting Future Water Needs in the Colorado Basin. For more than 60 years, since the 1922 Colorado River Compact, the river has been overallocated but underused. This has been especially true for the upper basin, where water users still do not take their full share of the water. Lower basin water users, the Southern Californians in particular, have been able to divert more than their allotted share of water for many years. Under these conditions careful regulation of the river and accurate accounting for water weren't terribly important. In most years the river had at least a moderate supply for everyone who wanted it. What did it matter if a few million acre-feet of water got lost in the banks at Glen Canyon? That was upper basin water, and no one in the upper basin really needed it all that much. In that respect it was lucky that the main volume of bank storage built up during a time of water surplus in the upper basin. It won't have to be subtracted in the future when demand increases.

Potential needs for water in the upper basin are much greater than the supply available to fill those needs. As population increases, domestic use of water will increase. People moving into an area need jobs, and the growth of industry often accompanies the growth of population. So industrial needs for water are added to domestic needs. Much uncultivated but irrigable land in the region would need a water supply to make it productive. Finally, very large reserves of fossil fuel may some day require water for their exploitation. Not all of these competing needs for water can be met with the limited supply available in the upper basin.

What are some of the competing uses for water likely to receive attention in the future? More people will require more water for domestic use, but probably not in enormous quantities. Light industry isn't apt to use much, either. Irrigation, on the other hand, would be a large water user, as would an industry based on exploitation of fossil fuels. In the end choices among alternatives will be based both on local and on national needs.

Some water-using projects could be dropped from consideration without damage to national needs. On a national basis, the total area of productive agricultural land is more than adequate to meet our needs for food and fiber. Scrubbing a proposed irrigation scheme would be painful to local interests but would not be harmful to the country as a whole. On the other hand, some energy resources that exist only in the upper basin may one day be essential to our national security.

Practically all of the oil shale deposits in the United States are in Colorado, Utah, and Wyoming—all in the upper basin. The oil equivalent in the shale deposits is more than 10 times the known U.S. reserves of liquid petroleum. Mining and refining oil shale on a large scale would require much water, and the Colorado River and its tributaries in the upper basin would be the main source of supply. If national needs ever require large-scale exploitation of oil shale, the upper basin will need all its allocated supply and then some. Therefore future emphasis in the Colorado River Basin is likely to be on water conservation, along with a continuing effort to measure and account for water flow and water loss throughout the system accurately.

Stretching the Supply Through Conservation. The United States has been blessed with an abundance of natural resources, which have provided the physical base for our high standard of living. Because we have had so much, we have tended to be wasteful, and conservation has been an idea often taught in school but seldom practiced by the majority of people—until the Arab oil boycott of 1973, which ended a long era of cheap energy. Then our national government pushed the panic button and proclaimed an "energy crisis." Draconian measures were proposed for rationing existing energy supplies. Elaborate conservation programs were set up to encourage people to insulate their houses, to use solar energy in place of electrical energy, to drive smaller, fuel-efficient cars, and so on. Crash programs were instituted to provide alternative sources of energy from coal and oil shale to replace the embargoed Arab oil.

In the end, two factors acting together solved much of the problem and eased the crisis mentality that prevailed in the mid-1970s. These factors were (1) the large amount of waste in our energy use, and (2) the market price of energy. The United States has always had a semifree-market economy, and this has proved to be the most efficient mechanism for allocating resources. So when utility bills began to rise and the price of gasoline doubled, then doubled again, people made individual choices in the marketplace. The result was a drastic lowering in the rate at which the use of electric power had been increasing for years in the past. Utility companies began to find that they had surplus generating capacity. They postponed or canceled construction of new generating plants, and they bought less oil and natural gas for power generation. Many began converting to coal as a fuel to replace oil. Motorists began to plan their trips more carefully, bought smaller cars, and made fewer trips to the gasoline pump. The result was that the use of gasoline began to decline, and before long less and less imported oil was needed, and the price of crude oil actually began to decline on the world market.

The moral of all this is that conservation through choice in a market economy is apt to be the fairest and most efficient way to allocate resources. In the Colorado River Basin, where water resources are limited, this means that competing uses are ultimately apt to drive up the cost of water and in so doing drive out the users that can't pay a higher price for water. Because, for example, oil has a higher monetary value than agricultural crops, operators of an oil-shale processing plant can afford to pay a much higher price for water than a farmer competing for the same water for irrigation. The same goes for cooling water for a large power plant. Several years ago a Southern California utility company planned to build a large nuclear power plant on the lower Colorado River near Blythe, California (south of Needles). To provide for cooling water from the river the company bought several thousand acres of irrigated farmland (in an established irrigation district) to have access to water for the plant. The plant was never built, but this transaction shows the kind of choice that will no doubt be made in the future as population and industrial development increase throughout the basin.[3]

Stretching Supplies through Better Accounting Methods. As water becomes more valuable (in a monetary sense), the engineers and hydrologists who operate the river system will be under pressure to improve overall efficiency in accounting for all the water in storage or moving through the system. Better accounting methods may enable hydrologists to pinpoint locations or conditions causing losses and thus partially control the losses. Some losses, such as bank storage, probably are uncontrollable, given the geology of reservoir sites, such as that at Lake Powell. But eventually it should be possible to better account for those losses and predict how much water will return from the banks when reservoir levels fall. Losses due to evaporation, however, may be partially controllable if one is willing to pay the price. It has already been shown that releasing water from the cold-water zone in Lake Powell lowers temperatures in the river and in the eastern part of Lake Mead, thus reducing evaporation losses from the lake. Water released from Lake Mead now is taken from the cold-water zone, as at Lake Powell. This allows a large heat build-up (from solar radiation) in surface water of the western part of the lake near Hoover Dam and allows a substantial quantity of water to be lost by evaporation. It has been estimated that if water could be released from the surface of Lake Mead, it alone might reduce evaporation by about 100,000 acre-feet (123,350,000 m^3) per year. Releasing cold water from Lake Powell and warm water from Lake Mead could thus conceivably add about 200,000 acre-feet (246,700,000 m^3) to the lower basin supply. Changing the water outlets to the power plant at Hoover Dam from deep to shallow water is no doubt feasible. The practicality of such a shift in discharge sites will depend on the ultimate need and price of the water this technique would save.

Other operational strategies to conserve water may be devised as better methods are found for improving measurements of water moving through the system. As the price of water rises, the U.S. Geological Survey and the Bureau of Reclamation may be able to afford more gaging stations on the

---

[3]The example of "market price" for water, as cited here, is used to try to explain in simple terms a complex social, legal, and political issue that has been the subject of thousands of pages of legal briefs during years of court and legislative hearings. The dams were built and are operated from federal funds and the wholesale price of water is relatively cheap. That price will not rise much in the future for obvious political reasons. Those who buy this cheap government water, however, may find it more advantageous to resell it at a higher price rather than using it for their original purpose; e.g., the irrigation district and farmers who sold land with water rights to the utility company. It is in this sense that the author assumes a higher "price" or "cost" for water in speculating on future water uses in the Colorado River Basin.

main rivers and on more of the tributaries. Stream gaging on the river and a few of the larger tributaries and discharge measurements at the dams are the only places where discharge is measured regularly. Differences between river discharge measurements above and below the reservoirs are apportioned between changes in reservoir storage, evaporation, bank storage, and errors in measurements. Assumptions of flow in ungaged tributaries may cause the inflow and outflow figures to be wrong and result in misleading assumptions about evaporation and bank storage. Gages on the Green, Colorado, and San Juan Rivers are the sources of inflow data for Lake Powell. These gages are well upstream from the lake, however, and may not always give accurate figures for inflow to the reservoir. In a similar way the Colorado is gaged only at the Grand Canyon gage in the reach between Lake Powell and Lake Mead. Hydrologists simply guess at inflow to the Colorado between the Grand Canyon and Lake Mead.

Errors in measurements and estimates have not been of much concern in the past. Some errors tended to cancel each other as water moved downstream through the system, and whatever the net error was, it was not a large proportion of the basin's total yield. As competition for water causes its monetary value to rise in the future, however, hydrologists will be under pressure to reduce errors wherever possible. Although uncertainties are inherent in hydrologic measurements (or estimates), increasing the frequency of observations and the number of measuring points should tend to reduce errors and improve accuracy in accounting for all the water passing through the system.

Consider the estimation of evaporation from Lake Mead. The USBR hydrologists estimate evaporation by means of a mass-transfer method (as at Lake Heffner, Chapter 3) that uses water temperature measurements made near Hoover Dam in the far western end of Lake Mead. These temperatures are almost always higher than water temperatures in the eastern part of the lake, which is influenced by the cold water from Lake Powell. Evaporation estimates now being made probably are therefore consistently higher than actual evaporation over the whole lake surface. This may be the reason that USBR sometimes reports higher-than-expected storage in Lake Mead. If one or more additional measuring points were established on the lake at strategic locations, accuracy of evaporation estimates probably would improve and storage in the lake would be known with more certainty.

More gaging stations, especially on the tributary streams, would doubtless improve operational efficiency throughout the entire river system. In a similar vein a denser network of precipitation gages and snow courses in the mountains where most of the runoff originates would help the USBR more efficiently operate the reservoirs during the spring snowmelt.

These and other improvements in the gathering of hydrologic data are likely to take place when the need for water and the price of water rise sufficiently to justify the costs. It is not clear at this time, however, just how much additional gaging and measuring are worthwhile. Without budgetary constraints it would be easy eventually to reach a point of diminishing returns. Trying to gage every dry canyon or arroyo that occasionally carries runoff would be as impossible as trying to measure precipitation from every thunderstorm that occurs in this vast arid domain. Scientific feasibility and engineering practicality must be balanced.

In perhaps no other region of the nation do hydrologists and engineers face a greater challenge to refine their measuring, regulating, and accounting techniques for the flow of water through a river system. More than half the people living in the West depend directly on the Colorado for all or part of their water supplies. Because this is one of the fastest growing regions in the country, undoubtedly demands will grow for all uses of water that multipurpose reservoirs are supposed to serve—domestic use, irrigation, hydroelectric power generation, water quality control, fish and wildlife, flood control—not to mention all the legal requirements of the Mexican treaty, the Colorado River Compact, and various court decisions.

Will there be enough water at the right place at the right time to meet all these needs? Probably not. It seems safe to predict more battles in the future over water both in the courts and in the legislatures. What the outcomes will be no one can now say. Regardless of who wins and who loses, hydrologic measurements will influence the final decisions.

9

Chemical Quality of Natural Water

The water pure that bids the thirsty live.
(Ellen Underwood)

No matter how clear and sparkling it may be, natural water is never absolutely pure. Dissolved constituents are always present in solution. In technical language the water is called the *solvent* (because it is the major constituent), and the constituents in solution are called *solutes.* Some of the ways water takes solutes into solution were described in Chapter 2, which showed how the mechanisms involved depend on the polar nature of the water molecule and the action of the hydrogen bond.

Even in the laboratory, totally pure water is rare, because even if you can prepare it, you have great difficulty keeping it uncontaminated. Gases from the air will begin to go into solution if the water is exposed to the atmosphere, but an even greater hazard to purity is the container holding the water. Because water is such a universal solvent, it will begin to dissolve material from the walls of almost any vessel that contains it. True, the amount of material is usually insignificant from a practical standpoint; water will stay in a glass bottle indefinitely. Nevertheless, water in a glass container will never be absolutely pure; it will always contain a tiny amount of silica dissolved from the glass.

All natural waters are solutions of some kind, with the ocean holding the most substances in solution. More than 70 of the natural chemical elements have been detected in sea water, and many other substances such as gases and organic compounds are also present. Scientists now think that the oceans formed from condensation of water in the earth's primitive atmosphere not long after formation of the earth itself. Composition of the early ocean is thought to be similar to today's ocean. It also appears that most of the substances dissolved in the ocean probably came from the depths of the earth through volcanic emanations as the primitive earth became layered into core, mantle, crust, hydrosphere, and atmosphere. Used this way, the term *hydrosphere* includes all waters of the earth, including water in transit in the atmosphere.

Recent explorations along some of the midocean ridges have documented present-day additions of mineral solutes in sea water from volcanic vents on the sea floor. While the sea floor has been forming and reforming during most of the earth's history (no current age measurements older than about 200 million years have been found for rocks at the sea bottom), the oceans themselves probably have existed essentially unchanged for at least 4,000 million years. Evidence for this can be found in the record of sedimentary rocks, which have occurred in every major geologic period from 3,800 million years ago to the present. The processes of weathering, erosion, and deposition necessary to cause these sedimentary deposits to form require existence of the water cycle as well as the atmosphere and ocean.

One reason for the relatively stable composition of the oceans over eons is the limit to solubility that characterizes each element and substance in solution in sea water. Every substance has a unique limit to its solubility in any solvent, water or otherwise. When a solution becomes *supersaturated* with a particular substance, that substance begins to precipitate and forms a solid phase in contact with the solution. This may be a reversible process, depending on temperature, pressure, acidity of the water, and so forth. For example, if you put a large amount of sugar into a pan of cold water, part of it may remain in the solid phase. When the water is heated on the stove, if all of the sugar goes into solution, then when the water is cooled again some sugar may precipitate out as a solid. Solutes in natural waters behave similarly.

As an example, consider the dissolution and precipitation of calcium carbonate ($CaCO_3$) in natural waters. Calcium carbonate is soluble in acidic (low pH) waters such as rainwater, which is usually slightly acidic because of dissolved CO_2 from the atmosphere. When it goes into solution, the calcium carbonate molecule breaks down into a calcium ion (Ca^{+2}) and a bicarbonate ion (HCO_3^-). Because minerals containing $CaCO_3$ are abundant in the soil and rock where rain penetrates, Ca^{+2} and HCO_3^- are common ions in both groundwater and surface water. In this way $CaCO_3$ is transported in solution by rivers from the land back to the sea.

Sea water is normally alkaline (pH above neutral), and when the neutral or slightly acidic river water enters the ocean, the new environment alters solubilities of several dissolved constituents. Because calcium carbonate is much less soluble in sea water than in river water, it begins to come out of solution in the sea. It may be precipitated biogenically to form shells or skeletons for living organisms or chemically to form lime deposits on the sea bottom. When the organisms die, their shells or skeletons may also supplement the accumulation of lime on the sea bottom, which in time may be buried beneath other sediments and turned into limestone. After much more time the geological processes of mountain building may raise the buried sediments above sea level and erosion may expose the limestone to the atmosphere. Rainwater will then begin the slow process of redissolving the lime, and rivers will return it in solution to the sea to restart the cycle.

Many other inorganic constituents contributing to the salinity of sea water have also taken part (and are still taking part) in similar "geochemical" cycles. Despite all this cycling and recycling, the chemical composition and average salinity of the oceans appear to have remained fairly constant during the last 2,000–3,000 million years of geologic time. It would seem appropriate, therefore, at this point to find out what the term "salinity" means as it is applied to water.

WHAT IS SALINITY?

The word *salinity* is used both in a general sense and as a technical term to describe the saltiness of water. As a technical term salinity was used originally in relation to sea water and was defined as the total of dissolved solids (TDS), expressed as grams of salt per kilogram of solution. When referring to fairly concentrated solutions, as in the ocean, the salinity was often expressed as parts per thousand or as a percentage by weight of salt to water. Thus, sea water with a TDS of about 35 grams of salt per kilogram is said to have a salinity of 35 parts per thousand or 3.5%.

Most natural waters are much more diluted than sea water, and there are more convenient ways to express their salinity. Chemists usually express results of their analyses in terms of milligrams per liter (mg/L). Occasionally you may also see the solute concentration of water expressed as grams per ton. Since a metric ton is equivalent to 1,000 kilograms, or one million grams, the expression "grams per ton" is also equivalent to parts per million (ppm). The sea water described earlier would thus have a salinity of 35,000 ppm.

For dilute solutions, mg/L is, for all practical purposes, equivalent to ppm. Either one is more convenient than the original "parts per thousand" for expressing salinity of dilute solutions. It is easier to write and to compare results in ppm or mg/L than in fractional parts per thousand or in fractions of 1%. As a result, most current literature on water quality gives solute concentrations as mg/L or ppm.

When Is Water Saline?

One could also ask "When is water fresh?" It depends on what you mean by *saline* and *fresh*. Water can be classified several ways based on chemical quality; probably the simplest is that proposed by the U.S. Geological Survey:

Total Dissolved Solids (ppm)	
Less than 1,000	Fresh
1,000–3,000	Slightly saline
3,000–10,000	Moderately saline
10,000–35,000	Very saline
More than 35,000	Brine

Some people use the term "brackish" for water from about 1,000–10,000 ppm TDS. Criteria for terms such as *fresh, brackish, saline,* and *salty*, when applied to water quality, are not universal. The terms depend on individual tastes and on the purpose for the water. One person's unacceptable saline water might be another's only drinking water.

The U.S. Public Health Service Drinking Water Standards recommend water with a maximum of 500 ppm TDS. Many people live where such fresh water is unobtainable; they may drink water containing more than 1,000 ppm, and in a few places more than 2,000 ppm TDS. When the salt concentration exceeds about 2,000 ppm, most people find it too mineralized for domestic use, although in a few arid regions (e.g., around the Sahara Desert) some people regularly drink water with more than 2,000 ppm of dissolved solids.

DISSOLVED SOLIDS

Because it is such a potent solvent, water either does or could hold in solution most solids and gases found in nature. For this reason natural waters are complex solutions, with the total number of solutes seldom if ever fully known. The typical analysis of water used for public supply, industrial purposes, or irrigation ordinarily includes determinations for only a few items out of the many dozen that could be determined. The main reason for this is cost. A complete determination of all the constituents that occur in trace amounts (just a fraction of 1 ppm) would cost a great deal for each analysis. In addition many laboratories that perform routine water analyses are not equipped to analyze for some of the rarer trace constituents. Minor constituents that have been shown to be important to human health or to certain agricultural crops may be determined by the analyst upon request, but you pay more (sometimes much more!) for these determinations in an analysis. For example, one analytical laboratory was charging $250.00 per sample in the late 1960s for determinations of methyl mercury, a constituent causing much controversy at that time. This was about 10 times the cost (at that time) of a routine water analysis. Obviously, then, in an analysis you must ask only for what you really need when evaluating a water source for a particular use. Fortunately, this is seldom a problem, because most routine water analyses contain everything needed to evaluate the usefulness of a given water. The major constituents are nearly always included, and certain minor constituents also will be included, depending on the completeness (and cost) of the analysis.

Major Constituents

Most of the dissolved material in natural waters consists of the same few solutes, whether in the sea, in rivers and lakes, or under the ground. Of course, relative concentrations of these solutes differ in the different waters, but the same few substances always seem to turn up in most chemical analyses of natural waters. If you were to select an analysis at random, you would be likely to find the following:

Cations	*Anions*
Sodium (Na^+)	Chloride (Cl^-)
Potassium (K^+)	Sulfate (SO_4^{-2})
Calcium (Ca^{+2})	Bicarbonate (HCO_3^-)
Magnesium (Mg^{+2})	

In a typical analysis of sea water, these seven ions would make up more than 99% of the total dissolved solids. And if the water were evaporated to dryness, the ions would combine to form a series of mineral salts, a fact well known to the ancients who sometimes obtained salt by evaporating sea water. Some of the more common salts are sodium chloride, potassium chloride, magnesium chloride, magnesium sulfate, calcium sulfate, and calcium carbonate. In sea water, sodium chloride—common table salt—makes up about 78% of the total dissolved solids, and the sodium cation (Na^+) and chloride anion (Cl^-) are therefore the most abundant ions in the sea.

Although both Na^+ and Cl^- are present in most waters, the major cation in most fresh waters is Ca^{+2} and the major anions are SO_4^{-2} and HCO_3^-. Two other constituents often listed among the major items in a freshwater analysis are silica and nitrogen, both of which occur only as minor constituents in sea water.

The silica (SiO_2) in fresh water comes from the chemical weathering of silicate minerals, which make up most of the rocks in the continental crust. Unlike the ionic constituents listed earlier, all of which carry an electrical charge, SiO_2 does not act in solution as a charged ion. For the most part silica remains in solution in rivers until they empty into the ocean. There, the depletion of silica seems to be due largely to the abundant marine organisms (mostly phytoplankton) that use silica to build their skeletons; this is probably why dissolved silica is a minor solute in sea water.

Because almost 80% of the atmosphere is nitrogen, it is not surprising to find this gas as a solute in natural waters. Like silica, nitrogen is taken up by organisms in the sea, and analyses of sea water show very small amounts of dissolved nitrogen as a solute. Nitrogen occurs widely in fresh water, where it is usually reported in analyses as nitrate ion (NO_3^-). It is found in rainwater as well as in groundwater and in surface runoff from the land.

Minor Constituents in Sea Water

If we surmise correctly that the oceans have existed continuously for most of geologic time and that the water cycle has operated for about the same period, then the world's rivers have been bringing dissolved mineral matter from the land to the sea for a very long time. Rivers draining geologic terranes under which different kinds of rocks lie bring different kinds of minor constituents, which all collect in the ocean reservoir. In addition, the ocean has received mineral solutes from volcanic emanations.

Some of these constituents follow a cyclical pattern—from the sea to marine sedimentary rocks to weathered and eroded rocks of the land to rivers and back to the sea. An example mentioned earlier for a major constituent is the cycle of calcium carbonate. Some minor constituents, once they reach the sea, seem to remain permanently in solution—gold, for example. Still others (which apparently occur in such minute quantities that they are missed in water analyses) are found only in the tissue or skeletons of marine organisms where the organisms have accumulated enough of the element to be detectable in an analysis. Some examples are the metals zirconium and platinum.

A listing of all the known minor or trace constituents in sea water would include a majority of the known chemical elements as well as many molecular complexes. One could probably assume that many of the trace elements in the sea have also at some time been solutes in surface water and groundwater on land; however, it is doubtful that any one river or spring has ever contained as solutes all of the minor constituents now in sea water.

GLOBAL CIRCULATION OF DISSOLVED SOLIDS

Salt in the Oceans

The world ocean is not only the main reservoir for water on the earth, it is also the chief repository for salt. Despite some large continental bodies of salt water (e.g., Utah's Great Salt Lake; the Dead Sea in Israel; the Caspian

Sea and Aral Sea in Russia; etc.) and the numerous salt beds interspersed with other sedimentary formations within the continents, most of the earth's salt is still in solution in the oceans. It has been estimated that if all the ocean water were evaporated, there would be left a deposit of salt large enough to cover the entire earth to a depth of about 100 feet (30 m).

A little salt leaves the ocean each year and travels with the water that continuously moves through the water cycle. Like the water, the salt eventually returns in solution to the sea. Salt in circulation moves first into the atmosphere as essentially dry particles, which then travel downwind toward the land.

Salt in the Atmosphere

Winds blowing across waves at sea create a spray of foamy water that begins to evaporate as it travels down the wind over the sea surface. As water evaporates, some of the dissolved salts remain as tiny particles floating in the air. They are so small and light that they are swept aloft by the wind along with the water vapor. These are the salt particles described as condensation and precipitation nuclei in Chapters 3 and 4. Their main constituent is sodium chloride, with lesser amounts of other major solutes in sea water. Chloride ion is a common constituent of rainwater, especially in regions near the coast. Some authorities have suggested that as much as two-thirds of the Cl^- in solution in surface waters of the land came from the ocean via the atmosphere.

In a few regions with no through-flowing drainage back to the sea, salt carried by the winds and rain may accumulate until water in lakes and in the underground becomes too salty to use. An example is the inland region of Western Australia, centered around Kalgoorlie. The region contains an old, almost-level landscape undisturbed by deep-seated geologic forces for millions of years. For untold centuries rainfall has contributed salt to ephemeral lakes that dry up in summer and to a static body of shallow groundwater. People who live there must either catch the rainwater as it falls or see it run into the salt lakes or seep down into the ground to join the salty and unusable groundwater. When the gold mining boom reached its height in the 1890s, a few enterprising men who brought in stills and sold fresh water distilled from salt lakes or shallow wells made more money than many of the gold miners. Today the region gets its fresh water from a pipeline that runs from a reservoir in the mountains east of Perth 400 miles (644 km) across the desert to Kalgoorlie. Here in Western Australia the cycle of mineral solutes has been temporarily short-circuited. Some time in the geologic future the land will rise, rivers will run out to sea again, and all that salt will rejoin the global circulation.

While sea salt is often a major constituent of rainwater, especially near the coast, there are other solutes as well. Dissolved gases such as nitrogen and carbon dioxide are present, along with solutes derived from smoke and dust particles in the air. Carbon dioxide is especially important because of its association with free hydrogen to form carbonic acid (H_2CO_3) in solution. This makes the rainwater slightly acidic and is responsible for much of the chemical weathering that breaks down minerals in the rocks, thereby adding more solutes to surface water and groundwater.

The chemical composition of precipitation depends largely on the gases and solid particles locally present in the atmosphere. Rain falling in open country away from the influence of civilization ordinarily contains about 10 ppm or less of dissolved solids. Downwind from heavily industralized areas, the concentration of dissolved material may be several times that amount, and the varieties of solutes in the rain will reflect the composition of waste products discharged by smokestacks, automobile exhausts, and so on. Thus, gaseous sulfur compounds issuing from power-plant and factory smokestacks are thought to be one source of sulfuric acid (H_2SO_4) in the so-called acid rain that plagues large areas of Scandinavia and the northeastern United States and Canada.

Salt in Surface Water

Just as the chemistry of rainwater reflects the atmospheric environment that produced the rain, so the chemical quality of surface and underground water reflects the land environments over which and through which the water flows. As you would expect, waters on the land have higher concentrations of salts than atmospheric waters but much lower concentrations than sea water. In the United States, the U.S. Geological Survey reports that in about 50% of the country the TDS of river water is less than 230 ppm; in about 90% of the country it is less than 900 ppm. In general the concentration of dissolved salts varies inversely with stream discharge, with higher salt concentrations occurring at low stream stages and lower, more dilute, concentrations at flood stage. In humid regions the maximum and minimum salt concentrations in a single stream may vary by a factor of 2 or 3, while in arid regions the total salt content of the water may be 10 times greater at low water than in time of flood.

Why should low water contain a higher concentration of dissolved salts? A review of the discussion on stream hydrographs in Chapter 8 may help answer that question. The hydrograph in Figure 8.20 shows the variation in stream flow caused by a storm that moved across the drainage basin. The first part of the chart (on the left side of the graph) shows a steady, slightly declining base flow in the stream just prior to the beginning of rainfall. Then, as storm runoff begins to reach the channel, stream flow increases rapidly, as shown by the steeply rising limb of the hydrograph. When stream levels reach their peak and begin to subside. the falling limb of the graph indicates that the channel is now carrying a decreasing flow as water drains from the land after the storm has passed. The stream slowly returns to its base flow, shown again on the far right of the graph. If you had taken frequent water samples from the stream during the storm and plotted their TDS on the same time base as the hydrograph, the result would have been a curve moving in the opposite direction of the hydrograph. Total dissolved solids would be at a minimum when stream flow was near its peak and would begin to rise as stream flow diminished. The curve would show the highest value for TDS when all water was coming from base flow. This base flow, then, must be carrying the higher salt concentration. Where does the base flow come from? It comes from groundwater seeping into the stream bed. So it turns out that most of the dissolved mineral matter in stream water comes from groundwater increment to stream flow.

Salt in Groundwater

Of all natural waters, groundwater has the most variable total salt concentration. This is not surprising, considering the solvent properties of water and the long time that most groundwater stays underground. Depending largely on the kinds of rocks making up the aquifers, the TDS of groundwater may range from less than 100 ppm to more than 300,000 ppm. These variations in water chemistry occur not only from place to place but also at different depths below the surface at a single locality. It is not unusual to find waters of distinctive chemical composition in aquifers lying at different depths in a single groundwater basin. In many basins shallow groundwater is of better quality than water from deeper formations. One of several notable exceptions to this is the Great Artesian Basin, where shallow aquifers frequently contain poor-quality water and where water quality generally improves with increasing depth of aquifers in the basin.

The type of aquifer, as well as its depth, may also influence water quality. Water from aquifers in igneous and metamorphic rocks generally is of good quality with low TDS. This would also include most aquifers in volcanic rocks. Aquifers in unconsolidated sediments and sedimentary rocks carry water with the widest range in TDS, from less than 100 ppm to saturated brines with several hundred thousand ppm of salt. Probably some of the best quality (lowest TDS) groundwater comes from shallow alluvial aquifers on river floodplains where the aquifers are continually being recharged with river water.

HOW QUALITY AFFECTS USE

The chemical quality of water has little to do with the way water moves around the water cycle, but it has much to do with the way people (and other living organisms) use water in their daily lives. To simply say the words *water quality* is to beg the question *quality for what?* Most living things need water to carry nutrients to their cells and to carry away waste products. For marine organisms, water is the very medium for all aspects of life. Land plants and animals need some water to survive, but too much is fatal. For all life water can carry poisons as well as nutrients, and the quality may be just as important as the quantity for survival. When Coleridge's Ancient Mariner said "Water, water everywhere,/Nor any drop to drink," he was voicing the complaint of every shipwrecked sailor adrift on the ocean. When a brewery claims its beer is good because it is made with "Rocky Mountain spring water," that isn't just advertising hype—the quality of water really does affect the way beer tastes.

The term *quality* is often used in a relative way; water is spoken of as being of "good" or "poor" quality for a particular use (e.g., making beer). Since pure water is always the same simple solvent, it is obvious that the quality of a given water must depend greatly on which solutes it holds. In addition to the major solutes discussed above, some other standard attributes of water affect its usefulness and are routinely measured and reported in chemical analyses. Some of these are discussed here, along with their importance in determining the usefulness of water in particular applications.

Total Dissolved Solids (TDS)

The TDS of water is probably the most-used criterion of its quality. That single number denotes whether the water is fresh or saline, and consequently whether you can expect to use it for drinking, irrigating crops, or whatever. The chemist determines total dissolved solids in two ways: by evaporating a water sample to dryness and weighing the solid residue, or by adding together the total weight of the major ions found in the chemical analysis. If all the major constituents have been determined, the two totals should be in close agreement. As a further check, the analyst usually determines the water's electrical conductivity, which is a function of total ion concentration in solution, and hence is an index of TDS.

Electrical Conductivity (EC)

Electrical conductivity (EC), called *specific electrical conductance* by the chemist, is a measure of the ease with which an electrical current will pass through a solution. As discussed briefly in Chapter 2, pure water is a very poor conductor; however, ions in solution will carry an electrical current; the more charged ions, the more conductive the solution.

Theoretically, EC should be directly related to the TDS of a solution. This is strictly true only for a simple solution of a single ionic solute (e.g., sodium chloride) in distilled water. For more complex solutions including natural waters, EC determinations, while not completely accurate in measuring the true value of TDS, do provide a useful indication of total solute concentration, and hence of overall water quality.

The advantage of the conductivity method for classifying waters is the ease with which it can be applied in the field. One technician using a simple instrument can make many EC determinations in a day at a fraction of the cost of a comparable number of chemical analyses. True, this method won't tell which ions are causing conductance in the water, but it is an inexpensive way to quickly classify a number of water sources and to decide which ones warrant the expense of a complete chemical analysis. Another way EC measurements are useful is in monitoring water sources for changes in quality after initial chemical analyses have been made.

Hardness

Hardness of water is characterized by the ease or difficulty one has in getting a lather from soap and by the mineral scale that forms when water is heated, as in a tea kettle or in a steam boiler. Although several possible solutes are known to cause hardness in water, the most common are the carbonates and bicarbonates of calcium and magnesium. An analyst customarily reports hardness as milligrams per liter (mg/L) of calcium carbonate. The following classification has been proposed by the U.S. Geological Survey, but it is only a guide to classifying waters and is not offered as a precise scale of hardness.

Hardness Range in mg/L of $CaCO_3$	*Water Quality*
0–60	Soft
61–120	Moderately hard
121–180	Hard
More than 180	Very hard

As there is no exact definition for determining "hard" or "soft" water, it must depend, at least in part, on what one is accustomed to. A water one person would call moderately hard might be considered relatively soft by another. Really hard waters, however, are unmistakable. If you've ever stayed in a motel supplied with hard water and tried to take a shower with the tiny bar of soap many motels supply their guests, you know how frustrating it is to see the soap disappear before you're properly lathered and bathed.

On the other hand, most people like water with a little hardness. It usually tastes better, and it washes the soap off when you bathe. Very soft water feels slick on your body, as though the soap were still there. Water with a little hardness, which is to say with calcium and magnesium ions, is favored for irrigation, too, as it tends to maintain higher soil permeability and allow infiltration to take place. The differential effects of calcium and sodium ions on clay structure were discussed in Chapter 5.

Where hard or moderately hard water is used as a public supply, it is frequently softened so that it will have less than 100 mg/L of hardness. The softening is done chemically by causing the calcium and magnesium ions in solution to be replaced by sodium ions. Sodium salts are much more soluble in water and don't cause trouble with soap or cause scales the way the calcium and magnesium compounds do. As most people who live where water is hard know, if the public supply isn't softened, the individual householder can do it by subscribing to a water-softening service.

Minor Constituents in Surface and Underground Waters

As mentioned previously, natural water contains many constituents in trace amounts, some of which appear routinely in water analyses and some of which are seldom analyzed for or reported. For the most part minor constituents are dissolved substances occurring in concentrations of less than 1.0 ppm.

Except for CO_2 and a few ions such as HCO_3^-, Cl^-, and SO_4^{-2} that may be cycled from the atmosphere, most of the dissolved mineral matter in surface water and groundwater comes from the chemical breakdown during weathering of the minerals that constitute the rocks of the continental crust. The example of silica was mentioned earlier. As different rocks have different minerals, so the rivers or groundwater draining various geologic terranes have a variety of minor constituents in solution.

Taken together, the minor constituents make up less than 1% of the dissolved solids in natural waters, whether on land or at sea. Yet their importance to life is out of all proportion to their abundance. Since life apparently originated in a water environment (probably in the sea), it is not surprising that living matter, from bacteria to humans, contains as essential components a number of the trace elements found in natural waters. Human beings, along with most warm-blooded animals, are known to require trace amounts of the following elements: arsenic, chromium, cobalt, copper, fluorine, iodine, iron, manganese, molybdenum, nickel, selenium, silicon, tin, vanadium, and zinc. Plants also require several of these same trace elements, along with others, such as boron. Some of these are reported routinely in water analyses, whereas others are determined by the analyst only on special request.

Two minor constituents that are usually reported in water analyses are iron and manganese. These two elements, weathering products derived from rocks in the earth's crust, are found in most natural waters. As stated previously, both iron and manganese are essential nutrients for humans and many other living things. When they occur in concentrations greater than 0.3 ppm for iron and 0.05 ppm for manganese, they can cause problems in public water supplies. They are easy to remove in a treatment plant, but if they are not removed, they tend to stain porcelain plumbing fixtures a rusty brown and to leave unsightly stains on laundry.

Sometimes a very small concentration (or lack thereof) of a minor constituent may be of great importance to human and animal health and to plant nutrition. After it was discovered in the 1930s that fluoride in drinking water could aid in preventing tooth decay, fluoride determination became routine in analyses of water for public supply. Very pure water supplies that lack even a trace of iodine have led to thyroid problems and the development of goiters. It is interesting that although trace elements are essential, just a little more than the required amount of some elements is toxic, inhibiting growth or in extreme cases even causing death. Zinc in trace amounts stimulates growth in many plants, but in larger amounts zinc poisons the soil to such an extent that plants won't grow at all. Boron is an essential trace element for all plants, but when it is present in irrigation water in concentrations of as little as 1 ppm it inhibits the formation of fruit on orange and lemon trees. Similar examples could be cited for other essential elements.

On the other hand, some trace elements in water are apparently not only not essential for living organisms, they may be quite harmful even in very small amounts. Lithium is an element seldom determined in routine irrigation-water analyses, but research in California has shown that as little as 0.05–0.10 ppm of lithium in irrigation water can be toxic to citrus and avocado trees and to red kidney beans.

Other trace elements may be causing damage to organisms and are still unknown because their effects haven't yet been discovered. This would seem to be a fertile field for future research. Our understanding of natural waters is still incomplete.

Appendix I

Map of California Showing Locations of Examples Discussed in Text

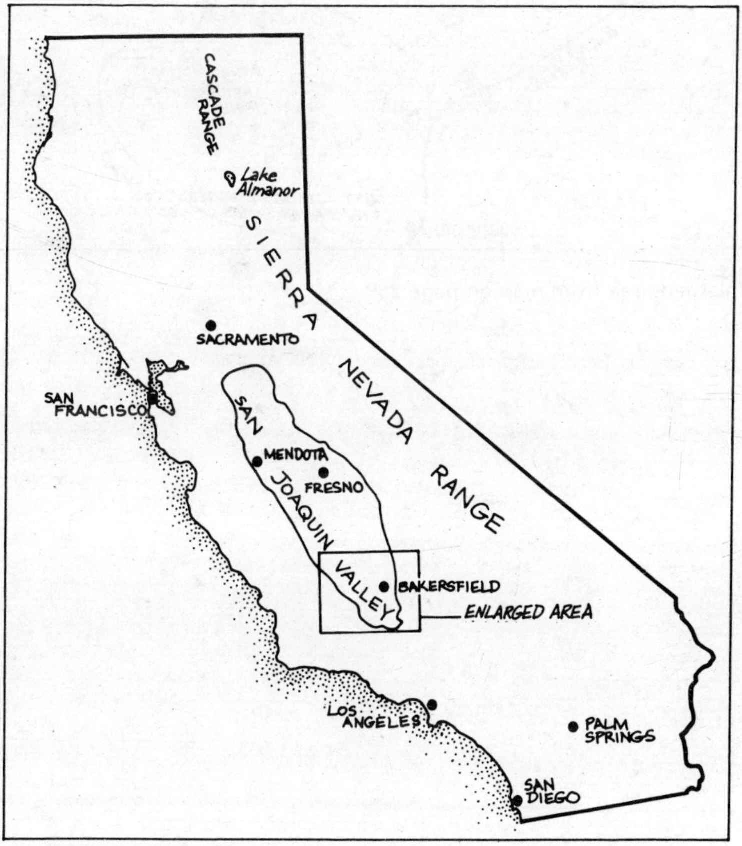

See enlarged area on page 260.

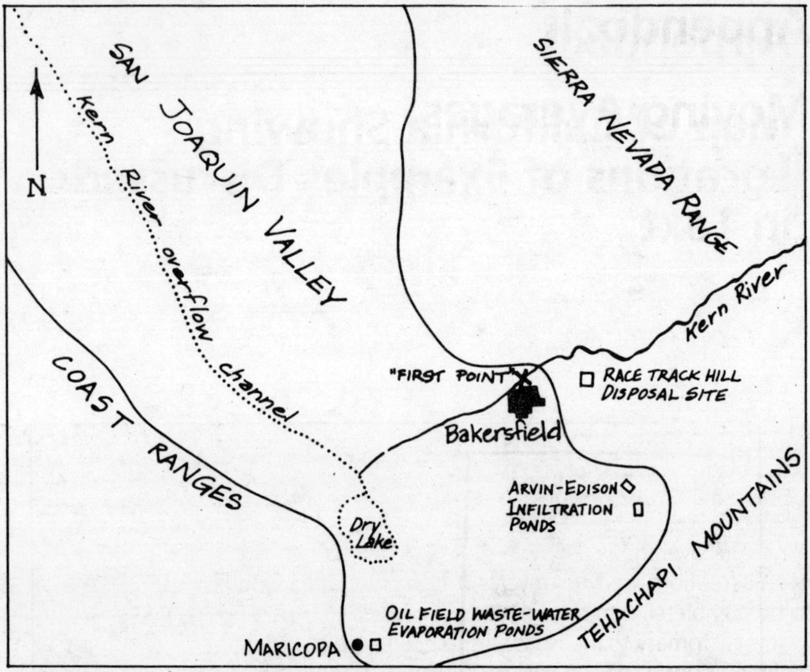

Enlarged area from map on page 259.

Appendix II

Moving Averages

A moving average is a simple statistical device that is useful for smoothing out fluctuations and for revealing trends in what would otherwise appear to be completely random data.

It is customary to measure and record hydrologic data at uniform intervals of time: hourly, daily, weekly, monthly, or annually. When the data are plotted on a graph with time along the horizontal axis, the result is what statisticians call a *time series.* An example is Figure 4.13, a diagram of annual rainfall at Bakersfield, California, for the 1874–1983 period.

Time series graphs are helpful in showing at a glance the range of values in the data for the period of record. Extreme values show up especially well on this kind of graph. Other characteristics of the data, however, may not be too apparent. A certain amount of statistical analysis may be required to bring out some of the more subtle aspects inherent in the raw data.

The simplest and most fundamental statistic, and the easiest to compute, is the arithmetic average. When this figure is computed for the period of record and plotted as a line across the graph (as in Figure 4.13), the data are separated into two groups, those occurring above and those below the long-term average. If there are any trends or cycles in the data, however, they still may not show up very clearly. One way to obtain more information from the raw data is to compute and plot a moving average.

A moving average fluctuates around the long-term average, and when the points are connected by a continuous line, a pattern of cycles sometimes shows up on the graph, as in the example of rainfall at Bakersfield. Sequences of "wet and "dry" years show up clearly on this graph. And while you could compute other statistical measures from the data and plot them on a graph, these two simple averages, plotted as shown here, are enough to provide a rough index of the regional climate and to allow comparisons with other sets of hydrologic data (see, for example, Figure 8.22).

Although moving averages can be computed for any desired time interval, the ones most often used for hydrologic data are the three-, five-, and nine-year moving averages. The five-year interval has been used for data in this book and is illustrated here. Methods of computation would be similar for averages at other intervals.

For computing the averages the data are listed chronologically from the earliest period to the latest, as shown in the accompanying table. The long-term average is computed (like any other arithmetic average) by adding all the values and dividing by the total number of items. The five-year moving average is computed by adding the first five items and obtaining their average value. Then the first item is dropped, and the sixth item is added, and another five-item average is computed. This process of dropping and adding items is repeated as the five-year period is worked down through the array of data. For convenience in computing and plotting, two columns are added to the data table listing the five-year totals and five-year averages. Books on statistics suggest placing the moving-average figure at the midpoint of the interval (in this case, opposite the third number in the interval), but because of the nature of hydrologic data it is customary in hydrology to plot the moving average at the end of each interval, as shown in Table II.1.

TABLE II.1 Precipitation at Bakersfield, July 1–June 30.

Season	Annual Total (inches)	Five-Year Total (inches)	Five-Year Average (inches)
1874–75	3.62		
76	5.00		
77	2.85		
78	7.98		
79	1.28	20.73	4.15
80	5.60	22.71	4.54
81	3.36	21.07	4.21
82	3.30	21.52	4.30
83	3.38	16.92	3.38
.			
.			
.			
.			
.			
.			
.			
.			
.			
.			
1974–75	6.71	29.49	5.90
76	4.37	27.21	5.44
77	4.19	28.40	5.68
78	11.73	32.11	6.42
79	6.69	33.69	6.74
80	6.84	33.82	6.76
81	4.78	34.23	6.85
82	6.22	36.26	7.25
83	10.33	34.86	6.97
Total:	633.81		

Long-term average: $\frac{633.81 \text{ inches}}{109 \text{ years}} = 5.81$ inches per year

Appendix III

Sources of Hydrologic Data

As mentioned previously, hydrologic data are collected throughout the country by both public agencies and by private observers. Whatever the source, most of these data find their way to a few central repositories where they are held in permanent storage for subsequent use by hydrologists, engineers, and others. Many of the data are ultimately released in published form, both in print and in computer data banks. Many of the publications, particularly those of the National Weather Service and the U.S. Geological Survey, are readily available in public and university libraries. If you don't find what you want in the library, you can address a letter of inquiry to one of the two agencies listed below.

For meteorological data such as precipitation, evaporation, air temperatures, humidity, and so on, send your inquiry to

National Climatic Center
Environmental Data and Information Service
NOAA, Federal Building
Asheville, North Carolina 28801
(704) 258-2850 Ext. 683

For data on river flow, floods, groundwater levels, water quality, water use, and so on, send your inquiry to

Hydrologic Information Unit
U.S. Geological Survey
420 National Center
Room 5-B-410
1220 Sunrise Valley Drive
Reston, Virginia 22092
(703) 860-6867

Many of the state governments also have water resources agencies, and sometimes you can contact them for local data that may not be in the federal files. In the author's experience, one of the most useful sources of information on government data files is the local political representative. Calling or writing to your local representative in the state legislature or in Congress and telling him or her precisely what you want is often the fastest and most effective way to get the information.

In countries outside the United States hydrologic data is most apt to be available from an agency of the national government.

Appendix IV

References

The following references are for the reader who wants to dig deeper into some topic discussed briefly in the preceding pages. Some are very elementary and some require a fair degree of technical sophistication, including advanced mathematics. Most have extensive lists of references on the subjects they cover.

A Primer on Water by L. B. Leopold and W. B. Langbein; U.S. Geological Survey, 1960, 50 pp.

This little book gives a simple, clear introduction to hydrology. It is written in nontechnical language and requires no knowledge of mathematics. First published in 1960, it has been reprinted several times and can be obtained from the U.S. Geological Survey. For information write to the USGS address listed in Appendix III.

Water: A Primer by L. B. Leopold; W. H. Freeman, 1974, 172 pp.

This book is an expansion of the USGS Primer on Water listed above, and like it, is nonmathematical in content. It is easy to read and understand and includes many good illustrations to help clarify the explanations given in the text. The chapters on *surface runoff* and on the *floodplain and the channel* are probably the best explanations of these topics obtainable anywhere in an elementary text.

Water: The 1955 Yearbook of Agriculture; U.S. Department of Agriculture, 751 pp.

Written for the layperson, this book contains a wealth of material on water and applications of water science to everyday life. There are 96 brief chapters grouped under such headings as *Where We Get Our Water, Caring For Our Watersheds, Water and Our Crops, Water and Our Wildlife, Pure Water for Farms and Cities,* and so forth. The chapters are authored by experts in the various fields and written in easy-to-understand, nontechnical language, without the use of mathematics. Although the book is obviously slanted toward agricultural topics and is 30 years old, it is an excellent source of basic information on many important aspects of water science.

Hydrology (Vol. 9 in Physics of The Earth Series, Prepared by a Committee of the U.S. National Research Council) edited by Oscar E. Meinzer; McGraw-Hill, 1942; reprinted by Dover, 1949, 712 pp.

Hydrology, the first U.S. book to offer a comprehensive description of the entire water cycle, has become a classic in its field. Although some of the analytical techniques described in the book are now outdated, the explanations of principles governing hydrologic processes are as valid today as they were when the text was written more than 40 years ago. As with any book authored by a number of experts, some chapters are more technical than others and some use a moderate amount of mathematics in explaining the principles or analytical procedures. The reader should not be put off by these occasional excursions into mathematical or technical language. Even the reader with no math background will gain much insight from the book, and it is recommended here as a logical starting point for anyone wishing to dig deeper into any topic under the general heading of hydrology.

Applied Hydrology by R. K. Linsley, M. A. Kohler, and J. L. H. Paulhus; McGraw-Hill, 1949, 689 pp.

As its preface states, this book is intended as a general reference for basic theory and methods of application. It is written primarily for use by engineers and its emphasis is on practical applications. Three revised editions have been published under the title *Hydrology for Engineers,* the latest of which is listed below. Anyone concerned with the latest analytical techniques as applied to current problems in engineering practice would probably pass up this book and go directly to the next one listed. However, this 1949 edition, like the 1942 edition of *Hydrology,* has become a classic in its field and should be consulted by any serious student of hydrology interested in gaining insight in the basic tenets of the science. If you don't understand the mathematical equations, ignore them and read ahead. The descriptive material is well-written and the maps and diagrams are excellent. Many of these were not repeated in subsequent editions, and this book is recommended here largely for the many maps and diagrams.

Hydrology for Engineers, Third Edition, by R. K. Linsley, M. A. Kohler, and J. L. H. Paulhus; McGraw-Hill, 1982, 508 pp.

This is primarily a textbook intended for engineering students with a fair amount of preparation in mathematics. It is a leading textbook in this field, particularly for surface-water hydrology, and is well worth looking at even if you don't understand the mathematics. It will give you an idea of kinds of problems that professional hydrologists are concerned with and the ways they work to solve those problems. For the student who may be interested in pursuing hydrology as a career, this book will indicate some of the skills needed for work in that profession.

A Primer on Ground Water by H. L. Baldwin and C. L. McGuinness; U.S. Geological Survey, 1963, 26 pp.

Like the *Primer on Water,* listed above, this little book gives a brief, non-technical discussion of groundwater in the United States. Because of its brevity many topics are not discussed, but the diagrams showing how water occurs in rocks and the map showing groundwater areas in the conterminous U.S. make the book a worthwhile acquisition for the beginning student of hydrology. For information on price and how to obtain the book write to the USGS address listed in Appendix III.

Basic Ground-Water Hydrology by R. C. Heath; U.S. Geological Survey Water-Supply Paper 2220, 1983, 84 pp.

This publication of the USGS was first published in 1983 and reprinted in 1984. Like the primers on water discussed elsewhere, this book will probably remain in print for many years. It is (in this author's opinion) the best description of groundwater hydrology available today. As Heath says in the preface, "It consists of 45 sections on the basic elements of groundwater hydrology, arranged in order from the most basic aspects of the subject through a discussion of methods used to determine the yield of aquifers to a discussion of common problems encountered in the operation of groundwater supplies." Some of the sections employ mathematical equations, but for the most part the equations are simple and a review section at the end of the book helps those who need to brush up on their knowledge of mathematical operations. By writing to the USGS at the address given in Appendix III you can find out the current price of the book and where to send for it.

Groundwater by R. A. Freeze and J. A. Cherry; Prentice-Hall, 1979, 604 pp.

This is an advanced textbook covering practically all aspects of the subject. The reader is assumed to be conversant with mathematical operations through elementary calculus and to be familiar with the elements of geology, physics, and chemistry as covered in beginning university courses. Readers possessing these qualifications will find this to be probably the best single reference on groundwater currently available. And the reader not so well prepared academically will find many well-written descriptions in plain English that, along with the many excellent illustrations, will help in learning about the technical aspects of groundwater hydrology.

Introduction to Soil Physics by D. Hillel; Academic Press, 1982, 364 pp.

This book offers descriptions of soil water, infiltration, and evaportranspiration that will be understandable to most readers, including those with only a limited background in science. The word "physics" in the title may tend to scare off potential readers who are a bit short on knowledge of physics and mathematics. Don't let it discourage you from browsing through the book. True, there is a fair amount of simple mathematics, especially in the sample problems (which, by the way, are worked out for you by the author). But most of the book is in clear, readable English with just the minimum of technical terms necessary for understanding the subject.

A Primer on Water Quality by H. A. Swenson and H. L. Baldwin; U.S. Geological Survey, 1965, 27 pp.

Like the other USGS primers on water listed above, this one is written in nontechnical language without mathematics. There is a short section on water's structure and unusual properties followed by discussions of "how nature affects water quality," "man and his wastes," "the characteristics of water" (including minerals in solution, hardness, biological considerations, etc.), "organics, dissolved gases and sediments," and "saline water." There are also some good illustrations to help explain the text. The diagram of how water quality is changed as water moves through the water cycle is especially good. This booklet, like the other primers on water, can be ordered from the U.S. Geological Survey. Write to the address in Appendix III for price and ordering information.

Study and Interpretation of the Chemical Characteristics of Natural Water, Third Edition, by J. D. Hem; U.S. Geological Survey Water Supply Paper 2254, 1985, 264 pp.

For the nonspecialist this is unquestionably the best book on the chemistry of natural waters available in English. Although the reader will need some knowledge of chemistry (about what you would get in a good high school course or first-year course in college), to get maximum benefit from the book, there is much of interest here for the reader who knows little more than the chemical symbols of the elements. An especially valuable feature of the book is the use of typical analyses to illustrate ranges of compositions of natural waters in various environments. It includes an extensive list of references for the literature on water chemistry.

Fundamentals of Meteorology by L. J. Battan; Prentice-Hall, 1984, 304 pp.

There are a number of elementary textbooks on meteorology currently available. This one was chosen for listing here because it is up to date, totally nonmathematical, and offers clear, very readable descriptions of meteorological processes with a minimum of technical terms. There is a complete glossary in an appendix for the technical terms used as well as a list of references for further reading in the subjects covered. The book is strongly recommended for the reader wishing to gain some insight into the atmospheric parts of the water cycle without getting lost in a maze of technical detail. Battan is one of the best at explaining complex subjects in simple terms without resorting to mathematical equations.

Harvesting The Clouds: Advances in Weather Modification by L. J. Battan; Doubleday, New York, 1969, 148 pp.

This little book provides one of the best available explanations of how clouds form and how rain and snow originate in the clouds. The processes and techniques for weather modification are explained in clear, nontechnical language without resorting to mathematical equations. In addition to ordinary "rain making" through cloud seeding, the author also discusses modifying clouds to suppress hail and lightning storms and the possibility of someday controlling hurricanes.

Index

Pages on which terms are defined are in **boldface;** illustrations shown by *